# QUANTUM WEIRDNESS

# QUANTUM WEIRDNESS

*William J. Mullin*

*Professor Emeritus, University of Massachusetts Amherst*

# UNIVERSITY PRESS

Great Clarendon Street, Oxford, OX2 6DP,
United Kingdom

Oxford University Press is a department of the University of Oxford.
It furthers the University's objective of excellence in research, scholarship,
and education by publishing worldwide. Oxford is a registered trade mark of
Oxford University Press in the UK and in certain other countries

© William J. Mullin 2017

The moral rights of the author have been asserted

First published 2017
First published in paperback 2020

Impression: 2

All rights reserved. No part of this publication may be reproduced, stored in
a retrieval system, or transmitted, in any form or by any means, without the
prior permission in writing of Oxford University Press, or as expressly permitted
by law, by licence or under terms agreed with the appropriate reprographics
rights organization. Enquiries concerning reproduction outside the scope of the
above should be sent to the Rights Department, Oxford University Press, at the
address above

You must not circulate this work in any other form
and you must impose this same condition on any acquirer

Published in the United States of America by Oxford University Press
198 Madison Avenue, New York, NY 10016, United States of America

British Library Cataloguing in Publication Data
Data available

Library of Congress Cataloging in Publication Data
Data available

ISBN 978–0–19–879513–1 (Hbk.)
ISBN 978–0–19–885436–4 (Pbk.)

Printed and bound by
CPI Group (UK) Ltd, Croydon, CR0 4YY

Links to third party websites are provided by Oxford in good faith and
for information only. Oxford disclaims any responsibility for the materials
contained in any third party website referenced in this work.

*For Sandra*

The Schrödinger cat figure on the cover was produced by researchers at the Austrian Academy of Science and the University of Vienna using "quantum imaging," in which a silhouette is formed by photons that were never near the object, while photons actually passing through the shape were never detected. The detected photons could be used to form the image because they carried, via interference, the same position and phase information as those actually "seeing" the object. See Chapter 10 in this book.

# Preface

Quantum mechanics is the science that is fundamental to understanding all nature. It explains the central workings of atoms, molecules and of the more complicated systems they form: solids, liquids, gases, plasmas, including biological systems. It has allowed us to penetrate the atom to see that protons and neutrons are made of quarks and gluons. We now know about the multitude of elementary particles, such as neutrinos and Higgs bosons.

What quantum mechanics tells us is that particles, for example, the electron or the helium atom, once thought to be small solid entities, often act more like waves than particles. In an opposite situation, light, once thought to be entirely explained by wave behavior, is found to have energy carried by particle like quanta called photons. There is a particle–wave duality in nature that was not fully realized before about 1920. Erwin Schrödinger developed his famous wave equation in 1925 and we have been using it ever since to explain the universe.

Waves do things that seem strange if you are expecting particle like behavior. If I drop a stone in a pond, a wave spreads out from the source and is soon all over the pond, while we would expect a particle to be more like the stone that caused the wave, centered in one place at a time following a well-defined path. There can be two or more waves on the pond at the same time, and they can interfere with each other. We will see that such a superposition means a particle can interfere with itself and basically be in many places at once. If you can identify a wave's wavelength, the distance between crests, then the wave must extend out some distance; if it is localized into only one crest, you are unable to identify a wavelength. In quantum mechanics, this is related to the famous Heisenberg uncertainty principle, in which simultaneous certainty about a particle's position and momentum is impossible. Waves can penetrate into places that are a bit unexpected. Quantum tunneling results in an electron more or less going right through the equivalent of a solid wall on occasion. On the other hand, particles often act, well, like particles: one often sees pictures of tracks from detectors in large accelerators and they seem to travel in well-defined lines there—very much as one would expect particles to behave. How does the wave interpretation apply there?

The predictions of the Schrödinger equation agree remarkably well with experiment, but the way we seemingly must interpret what it is telling us often seems very bizarre. The oddness is often described as "quantum weirdness." As another example, in one experiment two particles separate until they are very far apart. We make a measurement on particle 1 at a point A and get a result. Instantly, we are able to predict with certainty what the result will be on a similar experiment on particle 2 at point B. In classical mechanics, this is possible only if the particles possess built-in properties that are correlated before the measurement, and determine their results. But quantum mechanics seems to say that it was the measurement process that creates the results! Einstein called this "spooky action at a distance" and thought that a good theory should not include such nonlocality. But later experiments confirmed that this remarkable nonlocality seems a necessary component of the theory. We have been trying to puzzle out the weirdness ever since.

Quantum effects in nature can seem weird to us because we grow up seeing the world in the light of Newtonian mechanics, even if we have not studied physics at all. Balls follow smooth arcs, more or less like the particle tracks. We can certainly specify a baseball's position and speed simultaneously. Balls bounce off walls rather than tunneling through. So we become accustomed to this "classical" Newtonian way of viewing things; when atoms or electrons behave very differently, that is, quantum mechanically, we find surprises.

A deep understanding of quantum mechanics often requires some complicated advanced mathematics. One can say the words that interpret the equations, but without the math, one sees only half the picture. On the other hand, a lot of the basic ideas are accessible with just a minimum of algebra. Algebra alone does not give the whole picture, but it does yield considerably more depth of understanding than just words without *any* math. So the reader of this book should know basic algebra (and some trigonometry). An introductory classical (Newtonian) physics background is useful too, but we provide an appendix of physics terms to help those who have not had any physics or who have forgotten much of it. The algebra background is probably more important than the physics. There are four sections of the book in which the math manipulations, while still algebra, are somewhat more difficult than average. These are Secs. 12.1, 13.3, 14.1, and 15.4. If necessary, skip the math there, and get the basic idea from the words.

# Preface

This book is not a text in quantum mechanics. It does not solve the Schrödinger equation for any situation. While it discusses the nature of the wave functions in standard cases like the harmonic oscillator, the particle in a box, and the hydrogen atom, it does not tell how these are derived. Moreover its sampling of topics is more likely to be among those that satisfy the criteria of either illustrating the weirdness of quantum mechanics or some fundamental aspect of it.

The bibliography has a mixture of references to elementary treatments and full research books and papers in order to give credit to authors whose works were useful to me in writing this book and for readers who are at a more advanced level and want to pursue further information.

My thanks to Christopher Caron, a high-school-age friend whose interest in learning about quantum mechanics stimulated the writing of much of the material in this book. My long-time research collaborator Franck Laloë of École Normale Supérieure in Paris has for years helped spark my interest in quantum questions, and his advanced treatise *Do We Really Understand Quantum Mechanics* has been one inspiration for the present, much more elementary book. I thank Franck and UMass colleagues Robert Krotkov and Guy Blaylock for critical readings of the manuscript. However, errors and poor explanations are entirely my responsibility. Many thanks to copyeditor Elizabeth Farrell for making the text read more smoothly.

WJM
Amherst, MA

# Contents

1. Waves — 1
   - 1.1 Some history — 1
   - 1.2 Classical wave dynamics — 3
2. Quantum Particles and Waves — 10
   - 2.1 Planck, Einstein, deBroglie, and Heisenberg — 10
   - 2.2 The particle in a box — 12
   - 2.3 A note on probability — 15
3. Harmonic Oscillators — 16
   - 3.1 Classical oscillators — 16
   - 3.2 Quantum oscillators — 19
4. Superposition — 23
   - 4.1 The double oscillator — 23
   - 4.2 The two-slit experiment — 28
5. Entanglement — 32
   - 5.1 Two particles in the double well — 32
   - 5.2 A diversion on spin — 33
   - 5.3 One-spin superposition — 37
   - 5.4 Entanglement: Two-particle superpositions — 39
     - 5.4.1 "Spooky action at a distance" — 42
     - 5.4.2 What's a hidden variable? — 44
     - 5.4.3 EPR's argument — 46
6. The Mach–Zehnder Interferometer — 49
   - 6.1 Interferometers — 49
   - 6.2 Particle versus wave experiments — 50
   - 6.3 Delayed choice — 53
7. Bell's Theorem and the Mermin Machine — 56
   - 7.1 The Mermin machine — 56
   - 7.2 What quantum mechanics predicts — 57
   - 7.3 What a hidden-variables theory might say — 61
   - 7.4 A Bell inequality — 63

| | | |
|---|---|---|
| 8. | What Is a Wave Function? | 66 |
| | 8.1 Schrödinger's cat | 66 |
| | 8.2 Wigner's friend | 69 |
| | 8.3 Quantum interference of macroscopically distinct objects | 70 |
| | 8.4 Decoherence | 71 |
| | 8.5 Quantum Bayesianism | 77 |
| | 8.6 Tests of the reality of the wave function | 79 |
| 9. | Bose–Einstein Condensation and Superfluidity | 85 |
| | 9.1 Temperature | 86 |
| | 9.2 Fermions and bosons | 89 |
| | 9.3 The laser | 92 |
| | 9.4 Superfluid helium | 93 |
| | 9.5 Bose–Einstein condensation in dilute gases | 95 |
| 10. | The Quantum Zeno Effect | 101 |
| | 10.1 Measuring an atom's energy level | 101 |
| | 10.2 Rotating polarization | 104 |
| | 10.3 Bomb detection: The EV effect | 106 |
| 11. | Bosons and Fermions | 111 |
| | 11.1 Wave functions | 111 |
| | 11.2 Examples of Fermi and Bose effects | 112 |
| |    11.2.1 Average particle separation | 112 |
| |    11.2.2 The second virial coefficient | 114 |
| |    11.2.3 Interatomic forces | 114 |
| |    11.2.4 White dwarf stars | 115 |
| | 11.3 Inclusion of spin | 115 |
| |    11.3.1 Polarization methods | 116 |
| |    11.3.2 Transport processes | 118 |
| |    11.3.3 Ferromagnetism | 120 |
| | 11.4 The Hong–Ou–Mandel effect | 120 |
| | 11.5 The Hanbury Brown–Twiss experiment | 121 |
| | 11.6 What more? | 126 |
| 12. | The Quantum Eraser | 128 |
| | 12.1 Erasing with atoms, photons, and cavities | 130 |
| | 12.2 Using photons and polarization | 137 |

| | | |
|---|---|---|
| 13. | Virtual Particles and the Four Forces | 140 |
| | 13.1 Survey of the four forces | 140 |
| | 13.2 Sizes of the forces | 143 |
| | 13.3 Virtual particles | 144 |
| | 13.4 Photons and the electromagnetic force | 154 |
| | 13.5 Gravitons | 158 |
| | 13.6 Gluons and the strong interaction | 159 |
| | 13.7 The $W$ and $Z$ mesons and the weak interaction | 160 |
| | 13.8 The particle zoo | 162 |
| | 13.9 Forces in condensed matter physics | 163 |
| 14. | Teleportation of a Quantum State | 165 |
| | 14.1 How it is done | 165 |
| | 14.2 Measuring Bell states | 170 |
| 15. | Quantum Computing | 172 |
| | 15.1 Deutch's problem | 175 |
| | 15.2 Grover's search algorithm | 176 |
| | 15.3 Shor's period-finding algorithm | 176 |
| | 15.4 The solution to Deutch's problem | 177 |
| |     15.4.1 Preliminaries | 177 |
| |     15.4.2 The solution | 178 |
| | 15.5 Is anything built yet? | 180 |
| 16. | Weird Measurements | 181 |
| | 16.1 Weak measurement | 181 |
| | 16.2 Measuring two-slit trajectories | 182 |
| | 16.3 Measuring a wave function | 185 |

*Epilogue*     189
*Appendix: Classical Particle Dynamics*     191
*Bibliography*     197
*Index*     203

# 1

# Waves

> Just take your time—wave comes. Let the other guys go, catch another one.
>
> DUKE KAHANAMOKU

## 1.1 Some history

Quantum mechanics is the basic theory of matter. It was originally developed in the early part of the twentieth century to explain the properties of the atom, but it has gone much deeper and broader than that. It ranges from systematically explaining the properties of the elementary particles (quarks, electrons, protons, etc.) to giving a deep understanding of solids, liquids, and gases. It is the most successful scientific theory ever developed, allowing us to calculate details to many decimal places of accuracy.

The only known limitation of quantum mechanics is in joining it to the theory of gravity, that is, Einstein's general relativity theory; we still do not have a quantum theory of gravity. Nevertheless, quantum mechanics has still been remarkably successful in understanding many features of the cosmos, such as how the elements evolved in stars, and how black holes can decay. It is possible that a more fundamental approach, such as "string theory," will allow a synthesis of quantum mechanics and gravity. We will see.

In a first course in physics, we learn Newton's laws, $F = ma$, etc. These laws (and those of Maxwell, for radiation) tell us how to understand the motion of baseballs, satellites, and many properties of light and radiation. Such "classical mechanics" explains how large objects behave. But it does not seem to explain the properties of small objects, like atoms. Classical mechanics gives us the basic concepts of mass, momentum, energy, etc., that we need to interpret everyday life. We also carry those concepts over into quantum mechanics, but they are interpreted somewhat differently there. When we do an experiment on an atom, we must necessarily use large objects as our measurement devices (we have to have knobs we can turn with our hands), and there is a meeting of the large and the small, that is, of classical mechanics and quantum mechanics. But the assumption here is that quantum mechanics is the more fundamental theory and that it can, in principle, be applied to the large objects, so that we can derive Newton's laws from quantum mechanics. Moreover, there are instances where large-scale behavior (as, say, in the "superflow" of liquid helium at low temperature) defies Newton's laws, and only quantum mechanics can explain the experiment. Modern electronic equipment (based on the transistor) is based on our knowledge of solid-state physics from quantum mechanics. Nevertheless, we tend to look at nature from a classical mechanical point of view. When we throw a baseball, we have an intuitive understanding of where it will go. If it suddenly changed course by 90°, we would think that was weird. If the baseball seemed to be in two places at once, we might believe we were being tricked by a magician. But, in quantum mechanics, having a particle in two places at once is exactly what we are led by experiment to believe is true. Such weirdness is what we have to deal with.

As successful as quantum mechanics has been as a calculational device, there is a fundamental problem: interpreting just what it means about the nature of matter. The basic object in quantum mechanics is the wave function; we know it is to be used in a probabilistic way, but that still leaves open whether it is a "real" object or a kind of bookkeeping system existing just in our minds. We are led by nature into describing a kind of instantaneous wave function communication or correlation at arbitrary distances that would seem to violate Einstein's relativity principle that things cannot travel faster than the speed of light, and yet, well...no, it sneakily manages *not* violate it. The basic effects are just downright weird when we try to look at them from the point of view of classical mechanics. What quantum mechanics actually means has been debated for 90 years, and the debate continues even more energetically today. Most physicists in the past had not worried about the philosophical issues of interpretation; they left that to take place in esoteric journals like *Foundations of Physics* and took the attitude of "shut up and compute," since they could explain their experiments independently of the profound implications, which were left to the philosophers and a few physicists.

However, in recent years, experimenters in laser and atomic physics have come to be able to manipulate individual atoms, and the weird behavior has led to the possibility of quantum computing and similar applications that may mean the weirdness has practical applications! This has meant that the funny business in quantum mechanics has escaped from the esoteric journals and now is in the frontline physics journals, with even more people involved. Weirdness is now relevant, and even philosophical interpretation has become a hot subject.

In classical mechanics, a baseball, the moon, an atom, and an electron are all particles, while light is a wave. And yet, in quantum mechanics, an electron sometimes behaves like a wave and sometimes like a particle. The terms "particle" and "wave" are said to be complementary; the concept that applies depends on the experiment one is doing. The particle of light is the photon, which also has this dual behavior. The fact that an object can morph between wave and particle behavior adds to the weirdness when we try to view experiments from a classical mechanics point of view. *The amazing thing is that the mathematics of quantum mechanics can accommodate both behaviors and explain either kind of experiment; it is the interpretation we try to place on the mathematics that gets us into debates.*

The basic law of quantum mechanics is the Schrödinger equation, which, for a single particle in one dimension, looks like this (don't be alarmed!):

$$i\hbar \frac{\partial \psi(x,t)}{\partial t} = \left[ -\frac{\hbar^2}{2m} \nabla^2 + V(x) \right] \psi(x,t). \tag{1.1}$$

This beast is certainly one of the most fundamental equations of physics. Solving it is an exercise in advanced calculus, which is not the path these notes will follow. However, we *will* center our attention on the wave function $\psi(x,t)$ and how it is used and interpreted in quantum mechanics. We will avoid all of the advanced math and just use basic algebraic manipulations "at worst." The Schrödinger equation is what is called a wave equation; it has solutions that describe waves. So, we need to understand the basic properties of waves. But we also need particle concepts. So, we will spend some space going over the basic particle and wave concepts here.

## 1.2 Classical wave dynamics

We will use a few standard physics ideas, like momentum and energy, in our discussion. These originally arose in classical Newtonian mechanics. We have listed and briefly explained a number of these terms in the appendix at the end of the book. Just a qualitative understanding of these terms is probably all that is required for reading this book. However, in quantum mechanics, we solve a wave equation, and the wave function indeed acts like a wave, so we need a bit more detailed understanding of the physics of waves. That subject is usually treated in discussions of water, sound, or electromagnetism. Such ideas will be immediately used to understand matter waves. As a supplement to the rapid introduction to waves given in this chapter, I suggest the tutorial and animations given in the website associated with my book on sound: http://billmullin.com/sound/, which we reference occasionally below.

Common waves travel in an elastic medium, a material that has the ability to rebound from a displacement. Suppose we have a number of masses connected by springs. If you displace the first mass and let go, the nearest connected spring is stretched, which displaces the second mass, which stretches the second spring, etc. The potential and kinetic energies are carried down the chain in the form of a wave. To see an animation of this, go to this website associated with my sound textbook: http://billmullin.com/sound/AnimationPages/FigX-6.html. Similarly, if you drop a stone in water, the water is displaced downward, which pulls the neighboring water down, and a wave progresses outward from where the stone entered. Waves occur in many places: in the air, as sound waves; on guitar strings that have been plucked; and in the vacuum of space, as electromagnetic radiation such as, for example, light waves or x-rays. Here, we will concentrate on the waves on a stretched string (say, a bungee cord). I shake one end of the elastic cord and that causes a wave to travel down it. As in a water wave, the medium (a particular part of the string) moves up and down, but the wave itself (the elastic energy) travels along the string. So, we distinguish between *medium* motion and *wave* motion. If the wave motion is horizontal, the medium motion is up and down, that is, perpendicular to the wave motion; we call this a transverse wave. A useful animation, from the website that shows the distinction between wave motion and medium motion, is given at http://billmullin.com/sound/AnimationPages/FigsI-1&2.html.

There are several types of transverse traveling waves; we distinguish between impulsive and oscillatory waves. The former is a single or a series of bursts. (The corresponding sound wave would be formed by clapping one's hands.) It moves along as a localized bump on the string. An oscillatory wave has regular repeating pulses.

# Classical wave dynamics 5

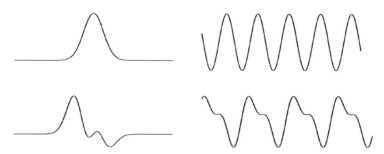

**Figure 1.1** Four wave types. The two on the left are pulse waves, while the two on the right are oscillatory (imagine the form continues to repeat beyond the figure). The top oscillatory wave is sinusoidal, but the lower one is non sinusoidal.

An important case of this is the sinusoidal wave, which has a special mathematical shape. Figure 1.1 illustrates an instantaneous picture of these waves. You have to imagine them moving along, say, to the right. The animation shows a sinusoidal wave in motion. A pulse wave in motion and reflecting from a wall can be seen at http://billmullin.com/sound/AnimationPages/FigsI-16&17.html.

It is possible to put two distinct waves on the same medium, but they will add to form a composite wave; this adding is known as *superposition*. (The second oscillatory wave in Fig. 1.1 is the superposition of two sinusoidal waves of different wavelengths [discussed later on in this section]). A very important property of waves is interference. Note that the transverse displacement of a wave can be positive or negative (above or below normal). If two waves pass through each other, where the two positive portions, or two negative portions, of the waves overlap, they will add, forming a larger displacement; but where a positive overlaps a negative portion, they will tend to cancel out. The reinforcement is called constructive interference, and the canceling is destructive interference. Take a look at the animation examples at http://billmullin.com/sound/AnimationPages/FigsI-19&20.html.

Waves reflect off boundaries; for example, sound wave reflecting from canyon walls result in an echo. In a concert hall, sounds "reverberate" by reflection, adding to the quality of the music. On a string (and elsewhere), the reflected wave can interfere with the original wave, producing what are called standing waves, which we talk about below.

Sinusoidal waves have a specific terminology; in the sinusoidal wave pictured in Fig. 1.1, the distance between the peaks (or the distance between the troughs) is the wavelength, denoted by the Greek letter lambda ($\lambda$); the time for any point of the medium to undergo a complete cycle (a round trip of some point on the string from the maximum through a minimum and back up) is called the period ($T$; measured in seconds). The frequency $f$ is the number of cycles per second (this used to be indicated as cps, but now the units are hertz [abbreviated as Hz]). Since the seconds per cycle is indicated by $T$, we have the relation

$$T = \frac{1}{f}. \qquad (1.2)$$

If the period is 0.5 s, the frequency is 2 Hz. The amplitude of the wave is the maximum distance any point is displaced from the normal string position. Thus, the distance from the minimum displacement to the maximum is twice the amplitude.

The distance a sinusoidal wave travels in a period $T$ is the wavelength $\lambda$. Thus, the wave velocity (distance per time) is given by

$$v = \frac{\lambda}{T} = f\lambda, \qquad (1.3)$$

a fundamental wave formula. The velocity of the wave $v$ is a property of the medium (depending on its elasticity, density, etc.) and is unchanging from wave to wave on the same medium. Thus, the shorter the wavelength of the wave, the higher its frequency, so the product is a constant.

As examples, consider sound and light: middle C in music has a frequency of 261.6 Hz, a wavelength of 1.32 m, and sound wave velocity of 345 m/s. Red light has a frequency of about $4.5 \times 10^{14}$ Hz, and a wavelength of $6.7 \times 10^{-7}$ m, with a wave velocity of $3.0 \times 10^8$ m/s.

Suppose we have a guitar string stretched tautly between two posts. Any wave on the string will be reflected from the ends, and the reflections will interfere with the original wave. The result is a wave that does not seem to move to the right or left but just up and down—a standing wave. An animation showing how two traveling sinusoidal waves form a standing wave is shown at http://billmullin.com/sound/AnimationPages/FigsII-2.html. Figure 1.2 shows a standing wave and its motion through one period. There are points at which there is always complete destructive interference; these are the nodes denoted by N in the figure. Positions at which maximum constructive interference occurs are the antinodes A in the figure.

# Classical wave dynamics 7

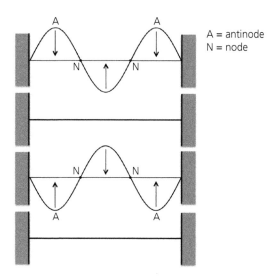

**Figure 1.2** A standing wave at times 0, $T/4$, $T/2$, $3T/4$, and $T$, where $T$ is the period of the wave.

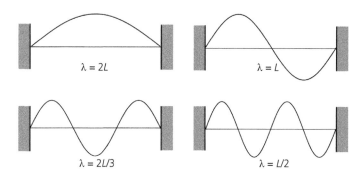

**Figure 1.3** Standing waves with wavelengths $2L$, $2L/3$, $L$, and $L/2$, where $L$ is the distance between the endpoints.

Standing waves can have only the wavelengths that "fit" into the distance between the walls, that is, have zero displacements at the ends. The four standing waves with the longest wavelengths are shown in Fig. 1.3. The wave with the wavelength $2L$ is called the fundamental wave. It has the frequency $f_1 = v/\lambda = v/2L$. The next longest wavelength is $L$, and that wave has the frequency $f_2 = v/L = 2f_1$. The next one has the frequency $f_3 = 3f_1$, and this continues. This sequence of frequencies is called the harmonic series. The frequency $f_1$ is the fundamental, or first, harmonic, $f_2$ is the second harmonic, etc. Not every frequency can occur on the string, only those that are multiples of the fundamental frequency. The spectrum of frequencies is discrete. (You might say it is *quantized*.)

When I strum or pluck a note on a guitar string, I rarely get a resulting wave that is purely only one of these harmonics; rather, I get a complicated wave that is a superposition of many harmonics. The sound resulting is pleasant because all the frequencies are in tone with one another; they are all multiples of the fundamental. Any arbitrary complex wave shape can be considered as a superposition of these selected harmonic standing waves. The nonsinusoidal oscillatory wave in Fig. 1.1 is made up of first and second harmonics. The sound excited by the string also has this property: it is composed of sinusoidal waves of these same frequency components. Figure 1.4 shows some complex wave forms of musical instruments. Each can be considered as being made up of a discrete set of harmonic sinusoidal waves; they differ in the relative amplitudes of the various components. When we analyze a wave into its harmonic sinusoidal components, we are doing what is called Fourier analysis. Any traveling repeating (oscillatory) wave can be constructed from a discrete harmonic sinusoidal series.

We can also do a Fourier analysis of an arbitrary wave shape, even if it is not repeating. For example, the pulse wave in Fig. 1.1 can be Fourier analyzed. However, the sinusoidal waves needed will not be a discrete

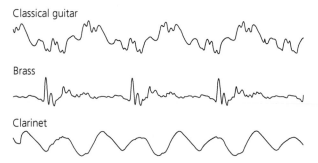

**Figure 1.4** Wave forms from selected musical instruments. What is plotted is pressure amplitude versus time. Our guitar string would produce a similar complex wave form. Each can be considered a superposition of sinusoidal harmonic waves.

set; rather, a set of sinusoidal waves with a continuous spectrum of frequencies will be needed—basically, every frequency, with the amplitude of each frequency depending on the specific wave. The pulse wave of the figure is peaked up over a narrow range of position. It vanishes outside that range. That means it takes a lot of sinusoidal waves that add up to make it exactly vanish everywhere except in a narrow range. The narrower the pulse wave is, the wider is the range of frequencies needed. If the width of the pulse is called $\Delta x$, and the range of frequencies is called $\Delta f$, then the relation is something like

$$\Delta x \frac{\Delta f}{v} \geq 1, \qquad (1.4)$$

where $\geq$ means "equal to or greater than." A small $\Delta x$ implies a large $\Delta f$, and vice versa. From Eq. (1.3), we can also write this as

$$\Delta x \Delta \left(\frac{1}{\lambda}\right) \geq 1. \qquad (1.5)$$

This last relation is closely related to the famous quantum Heisenberg uncertainty principle (see Sec. 2.1). We will see that the string's discrete spectrum is characteristic of allowed quantum frequencies too. Superposition and interference are central quantum ideas. Understanding general properties of waves is important to comprehending quantum mechanics. Water waves travel through water; sound waves travel through air. Light and other electromagnetic waves travel in a vacuum, so not really in any medium. The other strange wave is represented by the quantum wave function; the medium, if there is one, is not normal 3D space; it is expressed in what is called "configuration space," which makes is rather, well . . . weird.

# 2

# Quantum Particles and Waves

> The solution of the difficulty is that the two mental pictures ...—the one of the particles, the other of the waves—are both incomplete and have only the validity of analogies, which are accurate only in limiting cases.
>
> WERNER HEISENBERG

## 2.1 Planck, Einstein, deBroglie, and Heisenberg

We discussed briefly in Chapter 1 how quantum objects, like the electron or the atom, are expected to behave sometimes like particles and sometimes like waves. The two concepts, particle and wave, are complementary to one another. In some experiments, we get the expected particle-like behavior, and, in other experiments, the object behaves like a wave. A particle carries momentum $p$ (mass times velocity—for a particle with mass); a sinusoidal wave has a wavelength $\lambda$. Louis de Broglie, guided by an analogy with light, guessed that the relation between these two was given by

$$p = \frac{h}{\lambda}, \tag{2.1}$$

with $h$, the famous Planck's constant, to be discussed below. Clinton Davisson and Lester Germer in 1927 confirmed the de Broglie hypothesis by showing that electrons reflected from a crystal of nickel made a pattern (a diffraction pattern) on a screen that was entirely characteristic of waves. Schrödinger's wave equation for particles had been developed only a year or so earlier than this experiment.

The relation in Eq. (2.1) holds for electromagnetic radiation (light), too; light carries momentum (even though it has no mass). Max Planck and Albert Einstein showed early in the twentieth century that light is made up of quanta, indivisible packets of energy, called photons, that carry an energy proportional to the frequency of the associated light wave. The energy packet is

$$E = hf, \qquad (2.2)$$

where the size of the packet is measured by Planck's constant $h$. The photon energy is thus quantized. Einstein explained the photoelectric effect, in which electrons are ejected by light falling on the surface of a metal, by assuming this property. Max Planck had already made a similar assumption to help understand what is called "black-body radiation." In some experiments involving phenomena like diffraction and interference (we will see these later), light behaves like a wave, but in the experiments involving black-body radiation and the photoelectric effect, the particle behavior shows up. We can see that the two relations (2.1) and (2.2) are consistent: the shorter the wavelength, the higher the frequency, which implies large momentum and energy as well. Planck's constant is a primary measure of the magnitude of quantum effects.

Closely related is the Heisenberg uncertainty relation. Just as the particle and wave ideas are complementary, there are many pairs of physical features that are complementary to one another; an important pair is position and momentum. A quantum state can, for example, describe a particle that is quite localized in space, much like one of the waves on the left in Fig. 1.1. Such a particle will have a large spread in the possible values of its momentum, if that quantity is measured. If the quantum state specifies the atom's position to within an error $\Delta x$ (e.g., if we can say its position is between $x - \Delta x/2$ and $x + \Delta x/2$), then its uncertainty $\Delta p$ in our knowledge of what momentum $p$ would be found in a measurement satisfies

$$\Delta x \Delta p \geq h. \qquad (2.3)$$

If we put to $p = h/\lambda$ in this, we get $\Delta x \Delta \left(\frac{1}{\lambda}\right) \geq 1$, which is identical to Eq. (1.5), thus showing that the uncertainty principle comes from the wave nature of the particles. Note that, in the uncertainty limit on $\Delta x$ and $\Delta p$, we have not measured *both* $p$ and $x$. The principle is saying what quantum mechanics predicts about the uncertainties if one does *either* an $x$ measurement *or* a $p$ measurement.

However, there is another common way that the uncertainty principle is often interpreted, due to Heisenberg himself. Suppose I try to measure the position of an atom using a γ-ray (an energetic light ray that has a very short wavelength). This is like using a microscope to find out where a bacterium is. We can see where the particle is to some accuracy $\delta x$, but in doing so the collision between the γ-ray and the atom imparts a momentum $\delta p$ to the atom and so the final momentum becomes uncertain, due to the measurement process itself. In one particular situation, Heisenberg proved that $\delta x$ and $\delta p$ were related by Eq. (2.3), but in general this is *not true*, despite what a lot of textbooks imply. In fact, these two quantities are related by a different equation known as the *Ozawa uncertainty principle*,

$$\delta x \delta p + \delta x \Delta p + \Delta x \delta p \geq h, \qquad (2.4)$$

which makes it possible that $\delta x \delta p < h$ in some cases! Actually, there is a fundamental difficulty in even measuring the disturbance caused in momentum by a measurement in position or vice versa.

This section has skipped many fundamental experiments and theories constructed in the early part of the twentieth century. But we need to get on to the weird parts.

## 2.2 The particle in a box

The fundamental quantum quantity is the wave function. It describes what we can know about any quantum particle. The simplest quantum problem is a particle in a one-dimensional square box. Fig. 2.1 shows the system. The walls are impenetrable, that is, it would take an infinite energy to get through or over them, so that a particle just cannot go into those regions having $x < 0$, or $x > L$, at all; it is trapped inside the box. Within the box, it has zero potential energy (energy due to position) and constant kinetic energy (energy of motion). (See the appendix to review the ideas of kinetic and potential energies.) Some of the possible wave functions are also shown in the figure on the bottom. Because the box walls are impenetrable, the wave functions must vanish at the walls, and we must have a wavelength that just fits in the box. Thus, they look identical to the standing waves on a string, as shown in Fig. 1.3. Indeed, I just transcribed the figure. But the meaning of the curves is entirely different, as we will discuss. The energies of these states increase as the wavelength decreases, just as the frequency of the standing wave

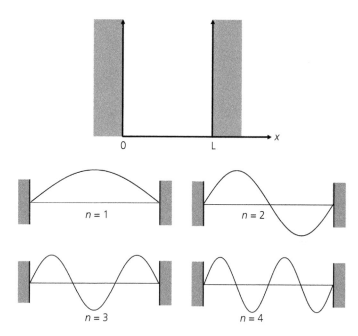

**Figure 2.1** A box potential well. *Top*: The particle sees a hard wall (infinite potential) at $x = 0$ and at $x = L$. So, it bounces off the wall without penetrating it. It feels no force while it is between 0 and $L$. The energy is always all kinetic. *Bottom*: quantum states of a particle trapped in the box. The states correspond to quantum numbers $n = 1, 2, 3$, and 4, with the number of nodes (zeros) being $n - 1$.

on a string increases as the wavelength decreases. The wavelengths are just $\lambda_n = 2L/n$, where $n = 1, 2, 3, \ldots$, as shown in the figure. The kinetic energy is $\frac{1}{2}mv^2 = p^2/2m$, so the allowed energies are (using Eq. [2.1])

$$E_n = \frac{p^2}{2m} = \frac{h^2}{2\lambda_n^2} = \frac{h^2}{8mL^2}n^2, \qquad (2.5)$$

where $h$ is again Planck's constant, and $m$ is the particle mass. The energies are quantized. The states with the shorter wavelength have greater energy. Our simple argument gives exactly the correct results we would get by solving the Schrödinger equation. Note that the lowest energy is *not* zero! This is an example of the uncertainty relation; because we know the position to within $L$ the length of a side of the box, the momentum must be uncertain, and a particle with momentum must have energy too. This lowest energy is called the *zero-point energy*.

Standing waves on a string can be thought of as being formed when a sinusoidal wave in one direction reflects, creating a sinusoidal wave in the opposite direction, so that the two interfere with one another to form the standing wave. Only certain wavelengths can do this. The waves shown in Fig. 2.1 can also be thought of as two waves of the same wavelength, that is, momentum, but in the two opposite directions, interfering with each other. The waves are required to have exactly the correct wavelengths to fit between the walls and vanish at the ends. We have momentum traveling in two directions simultaneously for each state. So the uncertainty in momentum is at least $2h/L$, with an uncertainty in the position of $L$, so the product is simply $2h$, in accord with the Heisenberg relation.

*The wave function*, call it $\psi(x)$ as is the tradition ($\psi$ is the Greek letter "psi"), *can be used to find the probability of finding the particle at each x*. But note that the wave functions shown are negative in some regions, and probabilities must always be positive! (See Sec. 2.3 on probability.) The probability of finding a particle at a particular place is proportional to the wave function *squared*, according to the interpretation of the wave function by Max Born (who came up with this idea in the 1920s), and this is positive.[1] Thus, for $n = 1$, you can see that the most probable place to find the particle is at $x = L/2$; the probability of finding the particle precisely at that position for $n = 2$ is zero, because there is a node there. The wave function has zero probability of being at positions outside the box; the wave function vanishes there and even at the boundaries of $x = 0$ and $L$.

Suppose we have a particle with no walls about, that is, with no potential energy curve, but just out in free space. Such a particle can have a precise momentum $p$ (with kinetic energy $p^2/2m$ and no potential energy), which is described by a sinusoidal wave of definite wavelength $\lambda = h/p$. Since there are no walls, any momentum is possible, in contrast to the case of a particle in a box potential. The momentum is precisely known, and the position is entirely unknown; the particle could be found anywhere.

---

[1] For readers familiar with complex numbers, we note that wave functions are often complex, so the probability is the absolute square. We will deal only with real cases until Chap. 16.

**Table 2.1** The probability of obtaining a particular roll when rolling a pair of dice

| Roll | Ways | No. of ways | Probability |
| --- | --- | --- | --- |
| 2 | (1,1) | 1 | 1/36 |
| 3 | (1,2) (2,1) | 2 | 1/18 |
| 4 | (1,3) (3,1) (2,2) | 3 | 1/12 |
| 5 | (1,4) (4,1) (2,3) (3,2) | 4 | 1/9 |
| 6 | (1,5) (5,1) (2,4) (4,2) (3,3) | 5 | 5/36 |
| 7 | (1,6) (6,1) (2,5) (5,2) (3,4) (4,3) | 6 | 1/6 |
| 8 | (2,6) (6,2) (3,5) (5,3) (4,4) | 5 | 5/36 |
| 9 | (3,6) (6,3) (4,5) (5,4) | 4 | 1/9 |
| 10 | (4,6) (6,4) (5,5) | 3 | 1/12 |
| 11 | (5,6) (6,5) | 2 | 1/18 |
| 12 | (6,6) | 1 | 1/36 |

The total of the fractions in the column on the right (the probabilities of each number showing up) is 1. A probability of 1 is a certainty; in this case it is certain that *some* number will show up when we roll dice. The roll with the highest probability is 7, since it occurs more ways than any other number.

## 2.3 A note on probability

The probability of a specific random event occurring is most often taken as the relative frequency of that event occurring in many instances of the same kind of event. If a particular random event A occurs on average every 3 out of 36 times the event happens, the probability of event A coming up again the next time is 3/36 = 1/12. Probability is always positive and between 0 and 1. Consider rolling a pair of dice. There is a single spot on one side, two on another, etc., up to six spots. I can roll a 3 in two ways: 2 + 1, or 1 + 2; there are a total 36 possible ways of rolling a pair. So the probability of rolling a 3 is 2/36 = 1/18. The other probabilities are calculated in Table 2.1.

# 3

# Harmonic Oscillators

> Human thought is like a monstrous pendulum; it keeps swinging from one extreme to the other.
>
> EUGENE FIELD

We have already introduced some of the weird effects that arise in quantum mechanics: matter waves, uncertainty, position probability, quantized energy levels, and zero-point energy. We will again see these in this chapter on the harmonic oscillator along with at least one new one: quantum tunneling.

## 3.1 Classical oscillators

Figure 3.1 shows a diagram (which is also shown in Fig. A.1) that illustrates how a swinging pendulum of constant total energy $E$ varies its type of energy between kinetic energy and potential energy. We want to treat the pendulum as a quantum particle and discuss how this simple classical system becomes a very interesting object in quantum mechanics; the problem illustrates many important features: the quantization of energy, the wave nature of particles, the wave function with its probability interpretation, quantum tunneling, the uncertainty relation, and the superposition of states. Moreover, when we have two oscillators acting together, a new feature, wave function entanglement, arises.

A pendulum is an example of a "simple harmonic oscillator." It is one of the simplest systems to treat in Newtonian mechanics or in quantum mechanics. (Actually, that statement is true only if the pendulum swings just a few degrees to either side. If the swing angle becomes large, then the problem of finding its properties becomes much more difficult. We will ignore that case!) The pendulum potential energy is a function of its position, that is, it is zero at the midpoint C and largest at the endpoints A and E, and similarly for the spring at the same locations. We can make a plot of this potential energy for the various position values

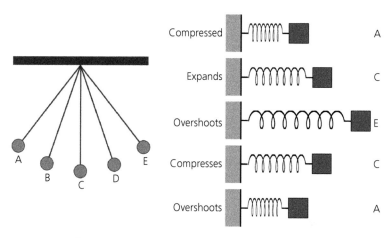

**Figure 3.1** *Left*: The pendulum has its maximum potential energy at points A and E, with its lowest at C. It has its largest kinetic energy at C, and its least at A and E. *Right*: Similarly, the potential energy of the spring is greatest when the spring is most compressed (A) or extended (E). The spring's kinetic energy is greatest at its midpoint of range (C).

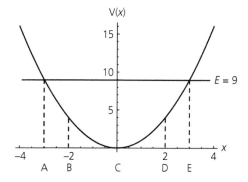

**Figure 3.2** A harmonic oscillator diagram showing the oscillator energy at various stages of its swing. At points A and E, all the energy is potential, while, at B and D, some of the energy is potential energy, and some of it is kinetic energy; at C all the energy is kinetic. Energy and position $x$ are given in units, but the units are not specified.

$x$. See Fig. 3.2. The position for the pendulum in this case is the angle of deviation from the vertical. We will still call it $x$. For the spring, C corresponds to $x = 0$, and A and E to the most negative and positive $x$ positions, respectively. In our figure, the total energy is $E = 9$. (Energy should be expressed in some units; joules is the common unit of energy, and distance x is usually expressed in meters, but we will not specify exactly, since the range of possible energies of oscillators in physics is so vast.)[1] The potential energy here is a quadratic function of position, that is, it goes up as the square of the stretch or deviation angle:

$$V(x) = \frac{1}{2}kx^2. \tag{3.1}$$

(For more details on the meaning of potential energy and this equation, see the discussion in the appendix.) It is traditional to call potential energy by the letter $V$, so we will do that here. In Fig. 3.2 I have, for simplicity, taken the "spring constant" $k = 2$ (units: energy/distance$^2$) so that $V(1) = 1$ but $V(2) = 2^2 = 4$, etc. (it is not "proportional" to $x$ but is quadratic in $x$). We take the total energy of the oscillator as $E = 9$ so that, at $x = 0$ (point C), the energy is all kinetic, $K = 9$, but, at $x = 3$ (point E), it is all potential, $V(3) = 9$. At $x = 2$, $V = 4$, so $K = 5$. The same numbers hold if $x$ is negative, which corresponds to the other direction of swing. In classical mechanics, where Newton's laws hold (that is, for large objects), the oscillator swings from A to E and back and cannot go beyond those extreme points (called "turning points"). If it did go beyond these points, the kinetic energy would have to become negative, which, of course, is impossible in Newtonian mechanics.

The potential energy is related to the force on the particle. Suppose the particle is at position A. On the curve, consider an arrow pointing down exactly along the curve (tangent to the curve). The angle of this "slope" (in calculus, this is related to the "derivative") is proportional to the force on the particle. The further up the curve you get, the steeper this slope and the greater the force. Wherever you are, the force always points back toward the center point C. (Mathematically, the force is $F = -kx$. For negative $x$ (on the left), it is positive, that is, it points right; for positive $x$, it is negative, that is, it points left.)

The pendulum, or the mass on a spring, oscillates back and forth with a set frequency $f$. A classical pendulum (or a classical mass on a spring) moves much more slowly near A and E, and faster at C. So, the most likely places to find it if you were to observe at random times would

---

[1] The Foucault pendulum in Paris has a 28 kg mass on a 67 m cable and so might swing with an energy of 300 J, but a rubidium atom trapped by a laser might oscillate with $10^{-31}$ J.

be at A and E, and the least likely place would be at C. We will compare this feature with the quantum pendulum below. You might imagine the potential energy curve as a real valley with a ball on the slope and gravity acting. The force is strongest on the ball at the highest points; starting the ball on the left, it would accelerate and move the fastest at the bottom of the valley, then slow down as it rose up the right side, then stop and slide back down, etc.

## 3.2 Quantum oscillators

Now let's consider a very small particle, like a neon atom, in a harmonic potential well. This situation can actually occur in nature. The neon atom can be in a solid held in place in an approximately harmonic potential well by the other atoms. (My first published paper was on just this subject.) There are also ways of creating a harmonic potential by using lasers so the atom oscillates around a low potential energy point; this is a very important field of study now. With quantum oscillators, we still have energy conservation; however, lots of things are very different. Now, if we fix the energy, we have trouble following the particle along. That is, we do not see it actually swing back and forth. We could examine its position, and if we look accurately enough, we could determine where it is at some moment, and doing so will certainly disturb its state afterward.

We use Schrödinger's equation to determine the allowed states and energies of the oscillator. We find that only certain states and energies are allowed. The energy is quantized.

Now let us draw the same kind of wave function curves for the pendulum, or the mass on a spring, as we did for the particle in the box. The allowed energies are shown in the graph of the potential energy $V(x)$ in Fig. 3.3(*Left*). The energies are

$$E_n = hf\left(n + \frac{1}{2}\right), \tag{3.2}$$

where $h$ is Planck's constant, and $f$ is the frequency of the classical pendulum back in Fig. 3.1, that is, its number of oscillations per second. (This relation looks much like that for the photon in Eq. (2.2). That is because there is a strong relation between harmonic oscillators and photons.) Note that our classical pendulum can have any

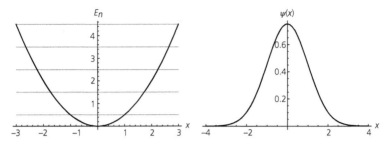

**Figure 3.3** Quantum pendulum states. *Left*: The allowed energies in the harmonic potential, $V(x) = (1/2)kx^2$, are shown as horizontal lines as given by Eq. (3.2). We take $k = 1$, and $hf = 1$, here in some energy units. These energies increase by one unit for each new higher energy state. *Right*: the wave function corresponding to the lowest energy of the oscillator, $n = 0$. This particular shape is called "Gaussian."

energy. In the classical oscillator, if you increase its amplitude (pull it out to a larger angle and let it go), you can increase its energy, and any reasonable energy is allowed. But, in the quantum oscillator, only certain energies are allowed. These are equally spaced: one energy unit $hf$ apart. Also note that while the classical pendulum has a lowest energy, $E = 0$, which corresponds to no motion at all, the quantum oscillator's lowest energy is not zero; there is a zero-point energy.[2] This zero-point energy again corresponds to the Heisenberg uncertainty principle; if I confine the particle to be trapped in the potential well, its position is known to an extent, and so it must have an uncertainty in momentum, and so its lowest energy is not zero. The wave function in Fig. 3.3(*Right*) could be expressed as a sum of many sine wave states (just like a pulse wave on a string) and so that corresponds to many wavelengths and so many momenta, by Eq. (2.1). The uncertainty in position is about the width of the wave function (say about three units in Fig. 3.3), and we could use that in the uncertainty principle to estimate the range of momenta this particle can have.

[2] Most physicists assume that quantum mechanics applies to objects of any size and that Newtonian mechanics can be derived from it. However, an oscillator the size of the Foucault pendulum has such a large energy and interacts so strongly with its surroundings that seeing the microscopic discreteness of the energy levels or the zero-point energy is impossible.

There are two other peculiarities of this wave function. The classical pendulum gets to points A or E and then turns around to go back toward C. It cannot get beyond these turning points. But the turning points for the lowest energy state, $n = 0$, are, as we see in Fig. 3.3(*Left*), at $x = -1$, and $x = +1$. But from the plot of the wave function in Fig. 3.3(*Right*), we see that the particle gets beyond those turning points and into regions that would correspond to negative kinetic energies in the classical case. The particle is not in any single place; in a sense, *it is in many places simultaneously* and has an average kinetic energy that is positive, even if some regions contribute negatively to the kinetic energy. This tail of the wave function has the particle doing a bit of quantum tunneling into classically forbidden regions. But that is what waves do—even light waves; they get into places where they are not allowed to run as free sinusoidal waves. As I said, the particle is in many places at once, and the wave function evaluated at a point $x$, when squared, gives the probability of being found at that particular point when we measure its position. The shape of the wave function in Fig. 3.3(*Right*) tells us that its most likely position is $x = 0$, as the center. That is very unlike the classical pendulum, which has highest preference for being at the turning points, as we mentioned above.

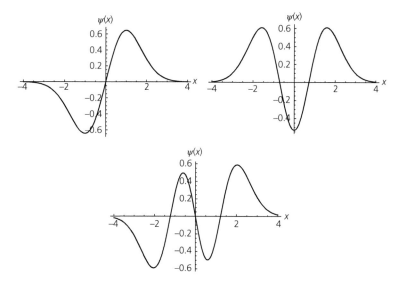

**Figure 3.4** Quantum pendulum states for $n = 1, 2$, and 3.

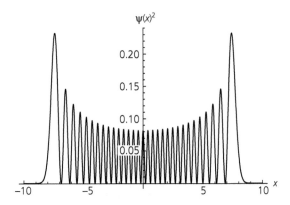

**Figure 3.5** Quantum pendulum probability function $\psi^2$ for $n = 30$. For this energy the most probable positions are near the turning points.

The next three energy states are shown in Fig. 3.4. Some regions of these wave functions have negative values. But we square the function to get the probability, which is then positive. The higher the energy, the more oscillations the wave function has. This is something like the waves on a string, where the higher frequency waves have more nodes. Again, these wave functions seem to prefer the central regions around $x = 0$ for the probability of finding the particle. However, if we go to very high-energy states, like $n = 30$, a state we see in Fig. 3.5, we find that the turning points become more favored. This corresponds to the usual feature of quantum mechanics that *we approach classical behavior when the quantum energy states become large*. This is an example of the correspondence principle of Niels Bohr: high quantum energy corresponds to classical behavior. In Chaps. 4 and 5 we use quantum oscillators to illustrate the concepts of superposition and entanglement.

# 4
# Superposition

> The greater the ambiguity, the greater the pleasure.
>
> MILAN KUNDERA

Waves add up. If two waves are on a medium at the same time, their net result is a sum of the two. When two waves have overlapping crests, we have constructive interference (adding), and where a crest and a trough overlap, we get destructive interference (canceling). We have seen a forward and reflected wave adding to form a standing wave. Superposition in quantum mechanics is the adding of two wave functions to form the total. Since wave functions have a probability interpretation, this feature can be basically having our particle in two distinct states at the same time. Here, we see the effect in the double oscillator and in the two-slit interference experiment. Superposition is the basis of an even more peculiar effect known as "entanglement," which we will get to in Chap. 5.

## 4.1 The double oscillator

What if we had the potential energy called a "double-well" potential, which is shown in Fig. 4.1? It is sort of hard to imagine a pendulum with this sort of potential energy. But we might try considering a particle in a slippery ditch shaped like this. Put it in the right side, drag it up the side to the right, and then let go. It slides down and then up the other side (there is no friction here). If you did not pull it very high up, then it won't get over the hump in the middle before stopping and sliding down and then back up the right side again. However, if you took it high enough (higher than the hump's top), then it would still have enough kinetic energy at the midpoint to go over the top, slide down to the left, up the side, back down, up over the hump, and so on. But, throughout this discussion, we are going to assume the particle has only a small amount of total energy—not enough to get over the hump with kinetic

24    Superposition

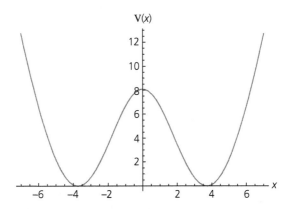

**Figure 4.1** A double-well potential energy curve. If a particle starts out in the right well, could it get to the left? Could it be in both at the same time?

energy left over at that middle point. We might draw a total energy line at, say one unit; since it would need eight units of total energy to get *over* the hump, it cannot. Note that if we put a classical particle with one energy unit in the right ditch, it will stay there. If we start it in the left ditch, it will stay there too.

But what happens to our *quantum* particle, which can tunnel into potential energy barriers? Since it might be able to tunnel *through* the barrier, it might be able to live on both sides simultaneously! If the barrier is very high, then the tunneling will not be deep enough into the barrier, and so if we start the particle on the right side, it will stay there, and the same if we start it on the left. The situation will be like the one shown in Fig. 4.2.

But suppose that the middle barrier is not so high, and, occasionally, if we were to start the particle out on the right, tunneling to the left side would be successful. This process leads to something a bit different than that described above. The Schrödinger equation, which we must solve to find the allowed wave functions and allowed energies, tells us that if the potential energy is symmetric, that is, looks the same on the right as the left, then the wave function must have symmetry too. (This is one of many uses of symmetry in physics; we can know something about the answer before we even find it.) There are two symmetries allowed: (1) the wave function looks the same on the right as it does on the left (it is symmetric), and (2) the wave function on the left is the negative of what it is on the right (it is "antisymmetric"). But we expect that, on

# The double oscillator

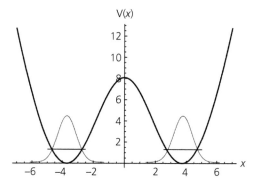

**Figure 4.2** A deep double-well potential energy curve with a very high barrier between the wells. The tunneling will not get deep enough in the barrier to take the particle into the other well. Thus, if it starts in one of the wells, it stays there. So, there are two equal-energy possibilities, depending on where it starts out. The energy value for each is shown as a short line. We consider only the lowest energy states.

the right side, it should look pretty much like what it did in the single well. So, we get answers that look like the graphs in Fig. 4.3.

The lowest energy state of the particle corresponds to the red (dashed, symmetric) curve, so the particle is most likely to settle down into that state. The green wave function (dash-dotted, antisymmetric) has an energy that is higher by a small amount, call it $D$, that depends on the size of the barrier, and the overlap of the two parts of the wave function. Each of these states has the particle in *both* wells simultaneously.

To get technical, write the wave function for the particle (call it particle number 1) in the left state as $\psi_L(1)$, and that for the right state as $\psi_R(1)$. Then, the red state (dashed) is very close to

$$\Psi_{\text{Sym}}(1) = \frac{1}{\sqrt{2}}\left[\psi_L(1) + \psi_R(1)\right], \quad (4.1)$$

with energy $E_0$, and the green state (dash-dotted) is about

$$\Psi_{\text{Antisym}}(1) = \frac{1}{\sqrt{2}}\left[\psi_L(1) - \psi_R(1)\right], \quad (4.2)$$

with energy $E_0 + D$. (Each of these states is multiplied by a factor $1/\sqrt{2}$, for a reason to be explained at the end of this section.)

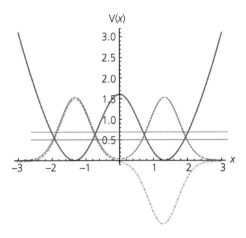

**Figure 4.3** A double-well potential energy curve where the barrier allows rapid tunneling. The tunneling allows the particle in one well to sneak into the other well. The two wave functions possibilities are shown in red (dashed) and in green (dash-dotted). The red one (symmetric) corresponds to a slightly lower energy than the green one (antisymmetric). There are higher energy states allowed, but we do not show these. If the lowest red energy is $E_0$, the green energy is $E_0 + D$; $D$ is often small.

How should we think about these wave functions? Say that our particle is in the lower energy (symmetric) state of Eq. (4.1). In the usual Copenhagen interpretation of quantum mechanics, it is not that the particle is in one well or the other and we just don't know which—there is just nothing more to know. It is in both wells equally. However, if we do a measurement, we will find the particle in one well or the other (if we measure accurately enough), but we cannot predict ahead of time which it will be. (The measurement process, in which the particle interacts with some detector [light perhaps] is said to *collapse* the wave function—in this case, perhaps to being in just one well.) This wave function property of being in multiple places or states at once is called "superposition"; we have superposed two very different positions here.

Suppose, however, we start the particle with small energy, so it is in these lowest states, but we know it is on the left side, that is, it is initially $\psi_L(1)$. (Perhaps we have just made a measurement that determined which side the particle was on and found the result $\psi_L(1)$.) We can describe this situation by simply adding the two wave functions of Eqs. (4.1) and (4.2):

$$\psi_L(1) = \frac{1}{\sqrt{2}} \left[ \Psi_{Sym}(1) + \Psi_{Antisym}(1) \right]. \tag{4.3}$$

The term $\psi_R(1)$ will cancel out completely, since it appears with opposite signs in the symmetric (red) and antisymmetric (green) functions. So the result is a particle completely on the left. But that state has a strange property. The new wave function is now a superposition of two states with different energies! The particle no longer has a fixed energy; it has two at once. But we should not be surprised that this can happen, since we said earlier that a particle in a box (a standing quantum wave) was made up of states of two momenta traveling in opposite directions and that a particle localized within a very narrow range of positions had a large uncertainty in its possible momenta. Now, it is energy that is uncertain.

If we follow how the combined wave function $\psi_L(1)$ varies in time, we will also see something interesting; it does not stay on the left but tunnels over to the right after a short time $T$ and then returns again to the left; it oscillates back and forth. If we know it was on the left at the beginning, we will be able to predict accurately that it will be found on the right after a time $T$. This is easily observed quantum tunneling. The state has a "mixture" of just two energy values, but neither one of them is big enough to allow it to go *over* the barrier; it just tunnels through regions where a classical particle would have negative kinetic energy.

The two energies are $E_0$ and $E_0 + D$, so the energy uncertainty is $D$. The time to tunnel, $T$, is found to have the value

$$T = h/D, \tag{4.4}$$

where $h$ is Planck's constant again. This is a form of a time–energy uncertainty principle. The higher the barrier, the longer the time $T$ and the smaller the energy difference $D$. When $D$ goes to zero (as in Fig. 4.2), $T$ is infinite, as we supposed there; no tunneling takes place.

You might think this is just abstract theory, but physicists frequently see this tunneling behavior; it explains many phenomena. One example is the molecule of ammonia, a model of which is shown in Fig. 4.4. The nitrogen tunnels between two equivalent positions in the molecule. There are many other examples of such quantum tunneling. For example, α-particles emitted from heavy radioactive nuclei are basically helium nuclei quantum tunneling out of the big nucleus. Helium atoms can form a solid at low temperature. A "vacancy" corresponds to

28                    Superposition

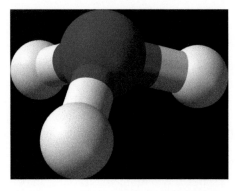

**Figure 4.4** A model of the ammonia molecule. The three white atoms represent hydrogens, and the blue, a nitrogen. The nitrogen sits above (or below) the level of the hydrogens and can tunnel between the two positions. The energy difference between symmetric and antisymmetric states is easily seen in radiation emitted or absorbed by the molecule. (From: https://en.wikipedia.org/wiki/Ammonia.)

a particle missing from the crystal. A neighboring atom can quantum tunnel into this hole. As subsequent particles tunnel into it, the vacancy forms a kind of quantum particle itself that moves in a wave like manner through the lattice. In Chap. 9, we will discuss very recent research on ultra-cold atoms, which can be seen tunneling, on an individual basis, through barriers formed by lasers.

Equations (4.1) and (4.2) each had a factor of $1/\sqrt{2}$ multiplying them. The factor multiplying each term in a superposition tells us the probability for the particle being in that state. To get the probability for one of the states, we square its factor. Thus, in Eq. (4.2), the probability of finding the particle in $\psi_L$ is $(1/\sqrt{2})^2 = 1/2$, and the probability of finding it in $\psi_R$ is $(-1/\sqrt{2})^2 = 1/2$. (So, what could the overall wave function be if the probability were 1/4 for finding it on the left, and 3/4 for finding it on the right?)[1] The midpoint of the right well in Fig. 4.3 is at position 1.25. The probability of finding the particle at this midpoint is proportional to $(1/2)\psi_R(1.25)^2$.

## 4.2 The two-slit experiment

Let's let L and R of Eqs. (4.1) and (4.2) stand for something quite different now. Consider the famous two-slit experiment, which is shown in

---

[1] One possibility is $(1/2)\psi_L(1) + (\sqrt{3}/2)\psi_R(1)$. Any others?

**Figure 4.5** The two-slit experiment. Particles from the source can go through either slit, and the wave function is a superposition of both possibilities. Note how the waves add or cancel by constructive or destructive interference, respectively, at the screen. When a particle is observed at the screen, it makes a single dot, but when all are added up, they show the interference between the two possible slit wave functions, producing bright and dark regions on the screen.

Fig. 4.5. In classical physics, this experiment was developed in 1803 by Thomas Young, to prove that light was a wave, as Newton and others had maintained that light was made up of particles. The wave interference in this experiment proved that light acts as a wave. Here we discuss the opposite; it will prove particles act as waves. Here, particle 1 (it could be an electron, photon, or other particle), when it gets to the screen, has the possibility of having gone through either slit: the one the left (L) or the one on the right (R). So, the wave function of particle 1 is

$$\Psi(1) = \frac{1}{\sqrt{2}} \left[ \psi_L(1) + \psi_R(1) \right]. \tag{4.5}$$

But the probability of finding the particle at the screen is the square of this. So, the probability of a particle being at some position $x$ on the screen is

$$P(x) = \frac{1}{2}\left(\psi_L(x) + \psi_R(x)\right)^2$$
$$= \frac{1}{2}\psi_L(x)^2 + \frac{1}{2}\psi_R(x)^2 + \psi_L(x)\psi_R(x). \quad (4.6)$$

The first two terms in Eq. (4.6) are squares and so they must be positive. Since wave functions can be positive or negative in places, the last term can be positive in some places and negative at others. Some places, it adds to the probability, and, some places, it subtracts from the probability, with constructive or destructive interference, respectively. It is this "interference term" that causes the variation of bright or dark lines shown in the figure. Water waves or classical light waves behave in this way. The amazing thing is that particles such as electrons or atoms do too.

There are some weird things going on here. Each particle going through the slits makes a small spot on the screen (it acts like a particle) as it hits a pixel on the detector screen; its wave function has been collapsed to a single point where the particle was detected. We get the interference pattern (a wave like effect) only when we look at the total effect of many particles having gone through the slits. But you might think that each particle went through one or the other and that it is the interference of one particle's wave with that of another that caused the pattern. That would be incorrect. We can slow the rate of particle flow through the slits so that only one particle is there at any one instant; we still get the pattern. We must conclude that *each particle interferes with itself*. That is, the probability distribution predicted by Eq. (4.6) applies to each particle independently of the rest.

Indeed, we can verify this claim. Suppose we put some device near each slit to determine whether the particle actually went through it. (It might be a weak light beam that would be interrupted by a particle passing.) Suppose the device tells us that the particle went through the left hole. Then, the wave function at the screen is just $\psi_L(x)$, and the probability is just $\psi_L(x)^2$. We keep doing this for all the particles; some will give $\psi_L(x)$, and some will give $\psi_R(x)$; the total result will be just

$$P_{\text{observed}}(x) = \frac{1}{2}\psi_L(x)^2 + \frac{1}{2}\psi_R(x)^2, \quad (4.7)$$

without the interference terms. (Note that 1/2 of the particles come from the left, and 1/2 from the right.) We have "collapsed the wave function" by looking. If we don't look at which slit the particle used, we get interference—a wave property. If we do look, we know the path the particle took—a particle like property. Particle and wave are complementary properties; the more one has of one, the less of the other there is—rather like knowledge of momentum and position.

Experiments equivalent to the two-slit experiment have been done with electrons, with photons, and even with carbon-60 and larger molecules, by Markus Arndt and colleagues at the University of Vienna. How big an object can we find that behaves in this way? Is there a limit so that an object bigger than a certain size no longer has this quantum property but must behave like a classical particle with no wave like aspects? Answering this question is a major goal of modern quantum research.

A basic assumption from the early days of quantum research was that there was a dividing line between classical objects and quantum objects. We needed measuring devices that behave classically (big magnets, cloud chambers, screens, dials) to make measurements and see the quantum behavior of the very small atoms, electrons, etc. This division was the basis of the Copenhagen interpretation of quantum mechanics and was developed mostly by Niels Bohr (who worked in Copenhagen). But we no longer really believe in this view. In principle, every object obeys quantum mechanics rules, but seeing the wave like behavior of a baseball is very difficult (it has a very small wavelength), and too many external influences are present that destroy its quantum behavior. As we learn to remove the effect of these outside influences, which is known as decoherence, we can see the wave behavior of larger and larger objects. And we have done this in some exciting cases, as we will see.

The fundamental questions involved with superposition and with seeing quantum behavior in large objects was made very clear by an example given by Schrödinger involving his cat. Schrödinger's cat has become one of the most famous (and misunderstood) examples (along with the uncertainty principle) of quantum behavior. We will examine it later.

# 5

# Entanglement

> From the outset, however, this whole controversy has been plagued by tacit assumptions, very often of a philosophical rather than a physical character.
>
> DAVID BOHM

## 5.1 Two particles in the double well

Let's go from superposition, which is the situation in which a quantum particle finds itself in multiple states simultaneously, such as two wells or two energy states, to an even more interesting case where two particles become *entangled* in two states. In entanglement, we don't know if Who is on first base or What is on second (See the famous Abbott and Costello comedy routine at http://www.youtube.com/watch?v=kTcRRaXV-fg) because they both are on both. Suppose we put two particles in the double well. We can do this in several ways. We could start both on the left (L1 L2) (L1 means particle 1 is on the left, etc.) or both on the right (R1 R2), or one on the left and the other on the right, but that could be (L1 R2) or (L2 R1). Using the names we wrote above, the (L1 L2) state, for example, would be

$$\Psi_{LL}(1,2) = \psi_L(1)\psi_L(2). \tag{5.1}$$

The RR state looks similar. Identical quantum particles have the feature that they are *indistinguishable*, that is, there is no way you can tell (L1 R2) from (L2 R1). (Well, there can be a way to "mark" them with some other internal property, perhaps, for example, with something called "spin," so one can tell the difference, but, without that, they are indistinguishable.) That means we would have to write

$$\Psi_{LR}(1,2) = \frac{1}{\sqrt{2}}\left[\psi_L(1)\psi_R(2) + \psi_R(1)\psi_L(2)\right]. \tag{5.2}$$

The state is a combination of the two particles, with particle 1 on the left, and particle 2 on the right, AND particle 1 on the right with particle 2 on the left. Could it be that one particle is really on the left, and the other on the right, but we just can't tell? No, it is more than that; the pair has both possibilities simultaneously. Einstein showed how this feature could seemingly cause contradictions that indicated that quantum mechanics was an incomplete description of nature. We need to discuss what he meant—however, we are not quite ready yet.

The factor $1/\sqrt{2}$ in Eq. (5.2) is necessary and tells us that the probability of each of the two states in Eq. (5.2) is the square of that quantity, namely, 1/2, as we have seen in other superposition cases above.

The state of Eq. (5.2) is said to be *entangled*, since we can't tell who is where. Entangled states are one of the most curious features of quantum mechanics. They allow us to design quantum computers; they give rise the possibility of teleportation and some really other peculiar effects where something seems to travel faster than the speed of light. Let's see what is so odd. To do so in the best way, we need to attach a different property to the particles. We could use photon polarization, to be discussed in Chaps. 6 and 12; instead, we will consider something very similar, namely, spin.

## 5.2 A diversion on spin

I am going to assume you know a bit about bar magnets; that they have north and south poles, that two north poles or two south poles repel, and that a north pole attracts a south pole. Electrical currents cause magnetism, as in an electromagnet or in permanent magnets, such as those made of iron, where the electrons themselves are little magnets, due to their own electrical currents. You could find some more information on magnets on the web, for example, http://www.howmagnetswork.com/. However, magnetism is really pretty complicated, and a full explanation would need a lot of background discussion, which we will avoid for now. (A full explanation of magnetic fields would allow us to investigate how magnets are used to guide and focus charged particle beams in particle accelerators such as the Large Hadron Collider at CERN, where the famous Higgs boson was recently discovered.)

Elementary particles have the property called *spin*. A proton, the nucleus of hydrogen, has what is called *spin one-half*, or spin 1/2.

Neutrons, quarks, electrons, etc., also have spin 1/2. Photons, gluons, and $W^+$, $W^-$, and $Z$ bosons have spin 1, and the Higgs boson has spin 0. Spin is somewhat analogous to the rotation of a planet spinning on its own axis, that is, rotating once per planet day. It involves angular momentum, and the amount of angular momentum for spin 1/2 is $\hbar/2$, where $\hbar$ is Planck's constant divided by $2\pi$. The earth, which obviously spins, also has an iron core, which causes it to be magnetic—it is a giant bar magnet with its magnetic south pole near the earth's geographic north pole. (This confusing situation is caused by the fact that a compass needle's north pole is attracted to and so points toward geographic north.) The elementary particles also have magnetism connected to their spin. Each is kind of a small bar magnet with a north and a south pole, which allows us to detect that the particle has spin. Indeed, the medical devices known as magnetic resonance imaging (MRI) machines are based on detecting the magnetic spin properties of the atoms and molecules in the human body, mostly the protons in water. The basic science using radiation to interact with spins is called nuclear magnetic resonance (NMR), which is one of my own subfield specialties.

We can measure this magnetism of a particle by passing it through an external magnetic field made by a large electromagnet or permanent magnet and which is not uniform, that is, it is weaker at the bottom and stronger at the top. Thus, if a little magnet has its north pole up so it sees a stronger magnetic field than the south pole sees, it will be dragged upward, while if it were oppositely directed it would be pulled down.

Suppose the bar magnet were sent through this large magnet exactly sideways; then, it would feel the same force on its north and south poles and not be deflected at all. If it were at a 45° angle, with its north pole tilted up, then it would be pulled only a bit more up than down. If it were aligned with the field, it would be pulled either strongly up or strongly down, depending on its direction. So, the degree of pull depends on the precise orientation of the magnet. But spin is different; if I send a spin through my vertical electromagnet, the spin is found either deflected *all the way up* or *all the way down*; it never has zero or partial deflection. The spin is quantized along the axis of my magnet. See Fig. 5.1, which shows the apparatus designed by Otto Stern and Walther Gerlach in the 1930s for measuring an atom's spin angular momentum; they used silver atoms but really measured

# A diversion on spin

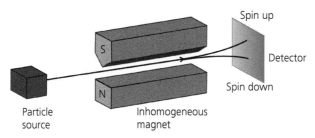

**Figure 5.1** The Stern–Gerlach apparatus. Spins go through the big magnet and are deflected by the nonuniform magnetic field. Classical magnets could be oriented in any direction, from pointing up to pointing down, with the up ones deflected upward, and the down ones deflected down. The ones pointing horizontally would not be deflected, and so the classical result would be smeared from up to down locations. But quantum spins are either deflected up or down along any direction and result in just two detector spots: one for up spins, and one for down spins; spin is quantized.

the spin of an electron in the atom. Also see the neat animation at http://en.wikipedia.org/wiki/Stern-Gerlach_experiment.

Let's look at a magnet in a nonuniform magnetic field in a bit more detail. We are going to use a coordinate system to specify directions $x, y,$ and $z$. See Fig. 5.2. We will show the magnetic field made by the Stern–Gerlach magnet as a series of lines. Where the lines are denser,

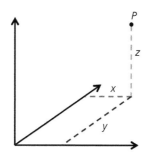

**Figure 5.2** An $xyz$ coordinate system; $z$ is up, and $x$ and $y$ are in the horizontal plane but are perpendicular to each other. One can specify any point $P$ in space relative to the origin by giving the numerical values of its coordinates $x, y,$ and $z$, as shown. In the Stern–Gerlach devices that we discuss, we assume the particle is traveling along $y$ (into the plane of the paper) and that the spin will be measured up or down along $x$ or $z$.

**Figure 5.3** *Left*: A small classical magnet with a north pole (n) and a south pole (s) sits at an angle in a uniform field caused by a large external magnet. The arrow indicates the direction of the small magnet. Because the field is uniform the small magnet is pulled neither up nor down along $z$. *Middle*: A small magnet in a *nonuniform* field where the field is stronger at the top will be pulled up. If it is flying into the magnetic field along a direction ($y$) pointed *into* the page it will take a curved path deflected upward. *Right*: Here, the small magnet is pointed down and will be pulled down in the nonuniform field.

the field is stronger. In Fig. 5.3(*Left*), the magnetic field is uniform, and a small magnet (like a compass needle) will not be pulled up or down, no matter what its orientation—whether straight up, straight down, or at an angle as shown. In the middle panel of Fig. 5.3, the small magnet is up along a magnetic field that is stronger at the top; this magnet is pulled up. In the panel on the right, the small magnet is pulled down by the large nonuniform field. In the Stern–Gerlach-type experiment, the particles would be shot into the large magnet on a trajectory that, in these figures, is into the page; the middle magnet would be deflected upward.

Now, consider a compass needle that is at an angle in the nonuniform field, as in Fig. 5.4. In this case, there is an upward force, but because of

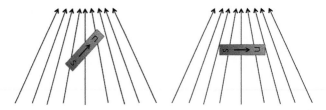

**Figure 5.4** *Left*: A small classical magnet sits at an angle in a nonuniform magnetic field. Because the field is not uniform, the small magnet will be pulled up along $z$. However, the upward pull will not be as strong as it was when it was along $z$, as in Fig. 5.3(*Middle*). *Right*: Here, the small magnet is pointed along $x$ (sideways), and the up and down pulls cancel out; no deflection would result when such a small classical magnet is sent through the field.

the angle, it is not as large as the case where the needle is straight up, so its deflection on being fired into the field will not be as large. The compass needle in Fig. 5.4(*Right*) is pointed horizontally along the x-axis and is equally much pulled up and down and so will not be deflected at all when it is shot into the field.

So, suppose we have a bunch of compass needles pointing in a large variety of directions, and we shoot them through the magnetic field. Some of the needles are up, some down, sideways, etc., and when we catch these on a screen as they come out of the field, each leaves a mark. We will have a smear of spots where they have landed. The highest mark will correspond to those that were up along the field (Fig. 5.3[*Middle*]) and the lowest to those that were down (Fig. 5.3[*Right*]), with those at any angle (Fig. 5.4) falling in-between.

But spin-1/2 particles are different. In measuring deflection along $z$, we get just two answers: all the way up or all the way down, with nothing in-between! Even when the particle goes in aligned along $x$, as in Fig. 5.4(*Right*), it is either found all the way up or all the way down. *Spin is quantized.* The probability of finding it up or down does depend on the angle at which it enters. The smaller the angle, the more likely it is that the particle will be found up along $z$. (A particle with spin 1 would have three possibilities: up, dead center, or down, with nothing in-between. Our compass needle is like a particle with spin one billion, say, with two billion possible deflections between the top and the bottom being available.)

## 5.3 One-spin superposition

We cannot visualize a real electron like a classical spinning ball of charge; it is a point particle (or a string, perhaps, in modern so-called string theory!), so spin is more mysterious than just being a spinning extended charge causing the electron to be a small magnet. But spin has all the properties of angular momentum. And we do find it quantized–it can be either up or down along any axis (like $z$) when we measure its deflection in the Stern–Gerlach magnet. We can say that the wave function of an up spin along the $z$ axis is $\psi_{\uparrow z}$, and that of one down along $z$ is $\psi_{\downarrow z}$.[1] If we shoot in a particle described initially by $\psi_{\uparrow z}$, it will surely be deflected

---

[1] The wave functions here represent only the spin directions. But, of course, the particle also has a probability of being at a position in xyz-space. This requires multiplying the *spin* wave function by another *space* wave function, as we show in Chap. 11.

upward by our SGA. The one described by $\psi_{\downarrow z}$ will surely be deflected down.

But what happens when we send in a spin that is not originally up or down? Suppose it goes shooting into the magnetic field, with the spin pointing sideways along x, as the classical spin did in Fig. 5.4(*Right*). (It is still moving along y [into the page], but the spin is pointing along x, and the magnet is stronger along z, so the deflection will be along z.) We know that we can get out only two deflections: up or down. If it goes in sideways along x, we might figure that it has a 50–50 chance of going up or down, and that would be correct. The way we describe that guess is by saying that the spin that is canted sideways as in Fig. 5.4(*Right*) is a sum of the upness and the downness along z. We write that as

$$\psi_{\text{up along } x} = \frac{1}{\sqrt{2}} \left( \psi_{\uparrow z} + \psi_{\downarrow z} \right). \tag{5.3}$$

Being up along x is equivalent to being in a superposition of equal amounts of being up along z and being down along z. (Note: again, the factor of $1/\sqrt{2}$ is needed, because of the probability interpretation of the wave function; when squared, the number in front of each term will give tell us the probability of each result. In this case, it is 1/2.)

Can you figure out how we might describe a spin that is at an angle, as in Fig. 5.4(*Left*), that is, when it is not fully perpendicular or parallel to z? If you can give just a rough idea of how this would go, you might be on the road to understanding superposition![2]

You could ask, how would I prepare a spin so that I knew it was up along x initially before sending it into my SGA aligned along z? The answer is to use two SGAs. The first has its nonuniform magnetic field aligned along x, and if the spin comes out deflected up along x, then I send it into the next SGA oriented to measure up or down along z. Also, I can block any spins that exit down along x from ever getting to the second SGA. Fig 5.5 shows this kind of double SGA.

Later, we will need the wave function for a particle that has its spin down along x; Eq. (5.3) was for a spin up along x; the down case looks

---

[2] If a spin is at an angle θ with the vertical, then a smaller θ means the spin is nearly aligned with z and is more probably found up along z. So, if the angle is, e.g., 45.0°, it turns out that one possible wave function is $\cos(45.0°/2)\psi_{\uparrow z} + \sin(45.0°/2)\psi_{\downarrow z}$; the probability of a spin-up measurement is $[\cos(22.5°)]^2 = 0.853$, and that of a spin-down measurement is $[\sin(22.5°)]^2 = 0.147$.

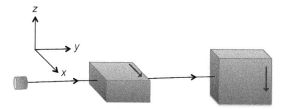

**Figure 5.5** Two successive Stern–Gerlach apparatuses, with particles traveling along $y$. The first apparatus, at the left, measures spin along the $x$-axis (it finds up along $x$, as indicated by the arrow). The second apparatus, at the right, measures along $z$ and finds down, as indicated by the arrow.

almost the same but with a minus sign in the middle:

$$\psi_{\text{down along } x} = \frac{1}{\sqrt{2}} \left( \psi_{\uparrow z} - \psi_{\downarrow z} \right). \tag{5.4}$$

This spin discussion can be adapted for use with photons, which can be in a superposition of vertical and horizontal polarizations, as we will see in Chap. 6. Considering spin in one direction as a superposition of spin in other directions is a remarkable idea. It leads to many weird quantum predictions, the first of which we now discuss.

## 5.4 Entanglement: Two-particle superpositions

We saw in Eqs. (5.1) and (5.2) how we might write a wave function for two particles in the double-well potential. But we can do the same kind of thing for spins. Suppose we had a positron (the electron antiparticle) and an electron; each have spin 1/2. If both have spin up along $z$, the wave function would be

$$\psi_{\text{up up } z} = e_{\uparrow z} p_{\uparrow z}, \tag{5.5}$$

where $e$ and $p$ stand for electron and positron wave functions. The wave function could involve motion as well as spin—with the electron, say, moving one way, and the positron, another. We might also have both with spin down along $z$:

$$\psi_{\text{down down } z} = e_{\downarrow z} p_{\downarrow z}. \tag{5.6}$$

In these two cases we see that the spins are pointing in the same directions, so we might expect they act like a single particle with spin = 1/2 + 1/2 = 1. Moreover, there is a third wave function that has total spin 1; it is

$$\psi_{\text{spin-1/up down}} = \frac{1}{\sqrt{2}} \left( e_{\uparrow z} p_{\downarrow z} + e_{\downarrow z} p_{\uparrow z} \right). \tag{5.7}$$

This is an entangled state that is a composite of up–down and down–up and is exactly analogous to the one in Eq. (5.2). The three states of Eqs. (5.5), (5.6) and (5.7) constitute a triplet of spin 1 states. If I sent them into an SGA-like device for charged particles, the first would predict the composite particle would be deflected up, the second down, and the third would have no deflection. These three states correspond, respectively, to having the total spin up along the field, down along the field, and with zero magnetic component relative to the field.

While the entangled state of Eq. (5.7) is useful and interesting, there is a much more mysterious up–down state. A neutral pion is a particle having spin 0; but it can decay into a pair consisting of an electron and a positron, both of which have spin. The final wave function of the electron–positron combination must certainly arrange the two spins to cancel out with one up and one down, but the result is this compromise:

$$\psi_{\text{spin 0/up down}} = \frac{1}{\sqrt{2}} \left( e_{\uparrow z} p_{\downarrow z} - e_{\downarrow z} p_{\uparrow z} \right), \tag{5.8}$$

with a minus sign instead of the plus sign of Eq. (5.7). This state is known as a "singlet" or spin-0 state. This state has an important property: it is rotationally symmetric—a term we will explain below; it is this feature that makes it special. (Again, the factor of $\frac{1}{\sqrt{2}}$ for each term in either Eq. (5.7) or Eq. (5.8) means that the probability of finding the first term in a measurement is $\left(\frac{1}{\sqrt{2}}\right)^2 = 1/2$, and the same for the second term.)

Now suppose our particles are in this singlet state, and the positron is moving at a fast speed to the left, and the electron rapidly to the right. We have two experimenters, Alice and Bob. (Physicists like to give names to the typical workers in the lab. Instead of calling them A and B, we introduce Alice and Bob. Other names also commonly appear, with Eve often trying to eavesdrop on the quantum-encoded message Alice is sending to Bob.) Alice is in position to measure the spin of the positron when it gets to her; Bob plans to measure the spin of the electron that is traveling toward him. If Alice finds the positron has up spin along $z$, then Bob must measure a down spin along $z$. The wave function was *collapsed* by Alice's measurement to the first term in Eq. (5.8). The electron and the positron might be a light year apart when Alice makes

# Entanglement: Two-particle superpositions

her measurement; it still instantaneously affects which result Bob will get. He always gets the opposite of her result.

Well, it might seem that this is no real conundrum; if I have a pair of gloves, one black and the other brown, and mix them up randomly in the dark, put one in one box, send that to Alice, put the other in the second box, and send it to Bob, then Alice has a 50–50 chance of finding a black glove, which immediately means Bob will find a brown one, and vice versa. So, there was equal probability of both possibilities; we just did not know which it was. But that is not what quantum mechanics seems to tell us. It says that both are simultaneously present at the same time; and neither one was the *true* one until the measurement was made.

The real weirdness comes from the symmetry of the above singlet state. Suppose the state was "prepared" in such a way that the spins were up and down along x; that is,

$$\psi_{\text{up down }x} = \frac{1}{\sqrt{2}} \left( e_{\uparrow x} p_{\downarrow x} - e_{\downarrow x} p_{\uparrow x} \right). \tag{5.9}$$

But, from Eqs. (5.3) and (5.4), we can do what is called a "change of basis," that is, we substitute for $e_{\uparrow x}$ by writing

$$e_{\uparrow x} = \frac{1}{\sqrt{2}} \left( e_{\uparrow z} + e_{\downarrow z} \right) \tag{5.10}$$

and for $e_{\downarrow x}$ with

$$e_{\downarrow x} = \frac{1}{\sqrt{2}} \left( e_{\uparrow z} - e_{\downarrow z} \right), \tag{5.11}$$

with just the same type of equations for $p_{\uparrow z}$ and $p_{\downarrow z}$. The substitution is

$$\psi_{\text{up down }x} = \frac{1}{\sqrt{2}} \left[ \frac{1}{\sqrt{2}} \left( e_{\uparrow z} + e_{\downarrow z} \right) \frac{1}{\sqrt{2}} \left( p_{\uparrow z} - p_{\downarrow z} \right) \right. \tag{5.12}$$

$$\left. - \frac{1}{\sqrt{2}} \left( e_{\uparrow z} - e_{\downarrow z} \right) \frac{1}{\sqrt{2}} \left( p_{\uparrow z} + p_{\downarrow z} \right) \right].$$

Multiply everything out, collect terms (the up–ups and the down–downs cancel out), and we get

$$\psi_{\text{up down }z} = \frac{-1}{\sqrt{2}} \left( e_{\uparrow z} p_{\downarrow z} - e_{\downarrow z} p_{\uparrow z} \right). \tag{5.13}$$

But this is precisely the same equation as in Eq. (5.8)! (The minus sign out front has no significance, since we square the wave functions to get probabilities. But, of course, the sign between the two terms *does* matter.) The up–down singlet state is up–down along *any* axis. This is also why we say the state is total spin 0; it is up–down along all directions. If one particle is found up along some axis, the other must be down along that *same axis*. All axes are equivalent. Well, the pion that produced them had no spin and was rotationally symmetric (no preferred direction in space), so the final state must be that as well. This is the heart of the mystery of this wave function.

### 5.4.1 *"Spooky action at a distance"*

So, in this peculiar state, Alice can choose to make her measurement along *any* axis she chooses; if she finds it up along that axis, then, instantaneously, Bob's particle must be down along the same axis, and vice versa. Alice and Bob can be very far apart when they make their measurements—so far apart that if their measurements are made at essentially the same time, a light pulse would not have time to travel the path back from Alice to tell Bob's particle what result that she found. Einstein in his theory of relativity showed that no signal could travel faster than the speed of light; but this experiment seems be "nonlocal" in that the result seems to violate relativity in having an instantaneous correlation between particles very far apart. Einstein did not approve of this "spooky action at a distance."

An important feature of spin measurements in different directions is that they satisfy uncertainty relations. Suppose Alice measures $\uparrow x$ so that, by Eq. (5.9), Bob's wave function is certainly $\downarrow x$, and the spin at his apparatus in the $z$-direction is completely uncertain, since, according to Eq. (5.4), it is an equal superposition of up and down along $z$! The variables $x$-spin and $z$-spin are, like position and momentum, "conjugate variables"; knowing one means the other is unknown.

The basic difficulty with "action at a distance" was treated in a famous 1935 paper by Einstein, Podolsky, and Rosen (collectively referred to as EPR). EPR make a fundamental hypothesis: "If, without in any way disturbing a system, we can predict with certainty (i.e., with probability equal to unity) the value of a physical quantity, then there exists an element of physical reality corresponding to this physical quantity." Given that Alice has made an $x$-spin measurement and gets $\uparrow x$, we can predict with certainty that Bob's spin measurement along $x$ will be $\downarrow x$;

and, according to EPR, there is then an "element of physical reality" corresponding to the $x$-spin at his apparatus. However, suppose Alice had originally chosen to measure $z$-spin and found $\uparrow z$; then, despite her far separation, according to Eq. (5.13), Bob's spin in that direction would now become certainly $\downarrow z$; and EPR tells us that his $z$-spin also has a corresponding "element of reality." Since EPR assume that Alice's choice could not affect Bob's result, they conclude that both $\downarrow$ $x$-spin results and $\downarrow$ $z$-spin results must have elements of reality simultaneously. But quantum mechanics also says that $x$- and $z$-spins satisfy an uncertainty principle and cannot be known simultaneously, so EPR conclude there is a contradiction and that quantum mechanics must be incomplete in some sense.

There is a way of making quantum mechanics "complete" and resolving this contradiction. One can suppose that there are some kinds of mechanisms ("elements of reality") within the electron and positron that correlate their spins, say, according to relative angle between the axis of measurement and some internal vectors in the particles. Such mechanisms have been invented that could make the correlation work as described so that each particle is simultaneously "ready" for any possible measurements. These are called "hidden" (or "additional") variables because the standard formalism of quantum mechanics does not describe any such features. EPR say that nonlocality is impossible since it seems to violate relativity; therefore quantum mechanics must be incomplete. Hidden variables might seem a way to make it complete to EPR's satisfaction. We give examples of possible hidden variables in Sec. 5.4.2.

Well, the story goes on to be even more interesting. Physicists have shown that one cannot send information by using the instantaneous correlations in quantum mechanics; so, in that sense, it does not violate relativity. All each experimenter sees is a random event, up or down along his or her own axis; Bob cannot tell from his measurements what angle Alice was using. Suppose Alice and Bob were both measuring along the $z$-axis. When Alice sees an up spin, Bob sees one down, and vice versa; but all Bob sees, after several measurements, is random up and down results. Suppose Alice now switches to measurements along $x$; when she sees her result Bob still gets random up or down results if he is measuring along $z$. He sees the correlations with Alice's angle *only* when they get together to compare their two sets of data.

Furthermore, in a rather amazing theorem, John Bell proved rigorously that any realistic theory (i.e., one using hidden variables) that is also local (any experimental result depends only on local settings, i.e., does not depend on any faster-than-light signaling from afar) leads to a mathematical condition that the data in a suitable experiment must satisfy. But quantum mechanics makes predictions that violate this condition! And real experiments agree with the quantum prediction. Bell's theorem (treated in Chap. 7) is perhaps one of the most remarkable proofs ever made in science or philosophy. It allows one to experimentally disprove the possibility of a whole class of physical theories of nature, namely, those that are local and realistic. As a result, physicists are often content with an interpretation of quantum mechanics that is non local (spin correlations at arbitrary distances), has no hidden variables, and/or is non deterministic (probabilistic).

### 5.4.2 What's a hidden variable?

Einstein suggested the usual view of quantum mechanics was incomplete because it did not eliminate the "spooky action at a distance." EPR suggested that there was something more needed to remove the apparent nonlocality from quantum mechanics. Hidden variables are possibly what EPR meant by the "elements of reality" mentioned in their paper. We can consider a model of how a hidden-variable theory might be constructed to explain the spin correlation in the electron–positron case. Ours is not a serious theory—it does not agree with experiment—but just a model to show how a hidden-variable theory might work. More precise models can be constructed. We suppose that there is a strange force on the spin that deflects the spin up or down with maximum deviation in a measurement according only to the hemisphere its magnetic north pole resides in. See Fig. 5.6. This differs from our descriptions in Figs. 5.3 and 5.4 of classical magnets for which the deflection up or down decreased as the angle got closer to 90°. In the model, the electron and positron are emitted along opposite directions. If the electron spin's alignment along any direction is positive (i.e., pointed in the upper hemisphere) it will be directed upward in a Stern–Gerlach experiment, with the same full force that it would have had if completely aligned; thus, it is found fully up when measured along that axis; if the alignment is negative, it will be directed completely downward and found down along that axis. The angle the electron makes with the $z$-axis is the hidden variable here. Correspondingly, the positron

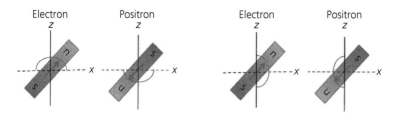

**Figure 5.6** A spin hidden-variable theory. If the electron's spin is measured along z, then it will be deflected fully up if it has its north pole anywhere above the dotted line, as indicated on the left; the positron will be found down along z if it is measured along that axis. As indicated on the right, in a measurement made along x, the electron shown will be found fully up along x if it has it north pole to the right of the solid line; the positron will have the opposite result. If the electron is found up along x, it could be at an angle between 0° and 180° (with respect to z), so that if the positron is measured along z, it will be up or down with equal probability, as found in experiments.

will have the opposite result. The measurements along x work out the same depending on whether the spin has alignment positively or negatively along x. Thus, the two spins are always oppositely correlated. There is no known magnetic force that would cause a full deviation (all up or all down) depending only on the sign of the component the spin had with an axis, so this model has no real validity, and this model's predictions do not agree with experimental results, but it shows how a hidden variable might operate. Quantum mechanics normally has no such additional descriptions of internal mechanisms to explain nonlocal correlations, so Einstein claimed it must be incomplete. But does quantum mechanics need additional descriptions? Its predictions already agree extraordinarily well with experiment without them.

Nevertheless, having hidden variables does not invalidate an interpretation of quantum mechanics. David Bohm was able to invent such an interpretation in which particles follow explicit paths (sort of like baseballs rather than waves) through a space that has a force field spread through it; it is a realistic theory. His mathematics is based on the Schrödinger equation, like all quantum mechanics, but interprets the mathematical quantities differently. The force field is described by the wave function. The theory is highly nonlocal; the force fields are set instantaneously through space and correlate one particle with another. In the two-slit experiment (where we get an interference pattern with one particle at a time going through the slits), a particle will go through only one slit or the other but will be guided by this force field to show an interference pattern on the screen after many particles have traveled through. Random initial starting points for the particles spread the particles over the possible classical paths. The particle trajectories are the hidden variables. If one, say, closed one of the slits, the force field would be changed everywhere instantaneously, guiding the particle to the open slit. The theory works, but most people do not feel it necessary to have such mental pictures of the underlying mathematics to explain what will happens. The mathematics of quantum mechanics does it for them without pushing it this far. The mantra of many physicists is simply, "Shut up and compute." But, lately, more physicists have not been shutting up as much and have been questioning the foundations of quantum mechanics.

We will have much more to say about hidden-variable theories in Chaps. 7 and 8.

### 5.4.3 EPR's argument

EPR's argument is different from the one in Sec. 5.4.1 and involving spin. (The singlet-state spin argument was invented later by David Bohm as a clearer, alternative approach.) But it is worth repeating the EPR argument to see their point again.

If we have two particles, we can consider two different uncertainty relations. Suppose the two particles, 1 and 2, have possible momenta $p_1$ and $p_2$, and positions $x_1$ and $x_2$, respectively. The particles each have the same mass. We can define the *average* position as

$$X = (x_1 + x_2)/2, \tag{5.14}$$

and the *total* momentum as

$$P = p_1 + p_2. \tag{5.15}$$

These two quantities obey the uncertainty relation

$$\Delta P \Delta X \geq h. \tag{5.16}$$

The uncertainty in total momentum times that in average position must be greater than Planck's constant. If you know one precisely, you cannot know the other precisely, as usual.

We can also define a *relative* position

$$x = x_1 - x_2, \tag{5.17}$$

that is, the distance between the particles, and a *relative* momentum

$$p = p_1 - p_2. \tag{5.18}$$

These also satisfy an uncertainty relation

$$\Delta p \Delta x \geq \hbar. \tag{5.19}$$

If I know one precisely, I do not know the other.

Suppose the two particles interact and then separate far enough that they no longer interact at all with one another. The basic assumption of locality here is that no matter what happens to one particle, it cannot affect the properties of the other particle, since they no longer interact. Perhaps they are a light year apart as well, but saying they do not interact anymore is all we need. Also, suppose the interaction has left them in a state where we know precisely both the total momentum and the relative position, that is, $P = P_0$, and $x = x_0$, where $P_0$ and $x_0$ are some very precise values, like 10 kilogram-meters per second and 5 meters, respectively. This is quite possible, since $P$ and $x$ do not satisfy any uncertainty relation together. (The terminology is that $P$ and $X$ are *conjugate* variables, but $P$ and $x$ are not conjugate.)

Now Alice decides to make a measurement on one of the particles, say, particle 1. She can decide to measure its momentum $p_1$ or she might decide to measure $x_1$ or, of course, something else. Suppose she measures $p_1$; then, immediately, we know that $p_2$ must be $P_0 - p_1$, by Eq. (5.15), and if Bob decides to measure momentum on his particle, that is what he must find. But suppose instead she decides to measure position of her particle and finds some specific value of $x_1$. Then, we know that Bob must find a value of position given by $x_2 = x_1 - x_0$ by Eq. (5.17). Alice's decision can be made at the last second before her measurement, with no way that information might be communicated to Bob's particle. But then, according to EPR, his particle must somehow have both pieces of information $x_2$ and $p_2$ built-in as elements of reality, since the locality assumption is that what happened to Alice's particle can have no effect on Bob's particle.

But quantum mechanics says that the wave function of Bob's particle cannot have certain information about both $x_2$ and $p_2$, since that would violate the uncertainty relation for particle 2. If $x_2$ is known, as it would be *if* Alice had measured position, then position cannot be known precisely, and vice versa. But her decision is made at the last instant and yet affects what Bob will find instantaneously. EPR argue that, given their locality assumption, there must be some further elements in the physics of Bob's particle (some hidden variables) that determines his $x_2$ and $p_2$ independently of what Alice does. Such a thing is not described in quantum mechanics, so the theory must be incomplete!

In quantum mechanics, Alice controls the way the wave function collapses by her measurement and that correlation is instantaneous whatever the distance apart the particles are. Quantum mechanics can be considered to be nonlocal. When she measures position of her particle, then the wave function of Bob's particle is one with certain position; when she measures momentum, then Bob's particle has a wave function with certain momentum. A key issue is the EPR statement that IF Alice's measurement was this, then Bob's result would be ..., OR IF the other measurement had happened, then Bob would have found .... A rule in quantum mechanics is that *unperformed experiments have no results*. Alice performs one or the other experiment; the one she did *not* do has no results; statements with "if" in them can get one in trouble. When you talk about what is real about a wave function, you must include the experiment that is to be done with that wave function. *It is the experiment that actualizes the reality.*

# 6

# The Mach–Zehnder Interferometer

> "Daddy, is it a wave or a particle?" "Yes." "Daddy, is the electron here or is it there?" "Yes." "Daddy, do scientists really know what they are talking about?" "Yes!"
>
> BANESH HOFFMANN

## 6.1 Interferometers

An interferometer is basically a variation on the two-slit apparatus discussed in Chap. 4. It can demonstrate the dual wave-particle behavior of photons (light quanta) or atoms. We will also use it to give an interesting example of entanglement. The apparatus we will study is known as a Mach–Zehnder interferometer (MZI) and is shown in Fig. 6.1. The input beam is split into two beams at beam splitter 1 (BS1). (A beam splitter for light can be a half-silvered mirror, which allows, on average, half the photons to be reflected and half to be transmitted; it can also be set to transmit more or less than half the photons). After the first beam splitter, the photon wave function becomes a superposition of wave functions existing in arm 1 and arm 2 of the interferometer:

$$\Psi_{\text{photon}} = \frac{1}{\sqrt{2}} [\psi_{\text{arm 1}} + \psi_{\text{arm 2}}]. \qquad (6.1)$$

The photons beams reflect at the mirrors and then interfere (showing wave behavior) at the second beam splitter (BS2), depending on the relative phases of the two waves (i.e., how the troughs line up with the crests). The interferometer can be arranged so that all the counts occur in detector D1 and none in D2 (or in any other desired count ratio). In other words, the interference is fully constructive on the path to D1 and destructive on the path to D2.

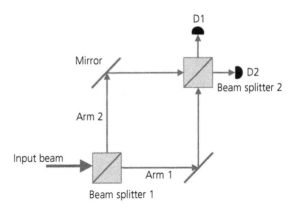

**Figure 6.1** The Mach–Zehnder interferometer. A photon beam entering at the lower left is divided into equal components at beam splitter 1, and these are deflected by mirrors to meet again at beam splitter 2. Here, the two beams interfere to produce counts in the two detectors, D1 and D2.

## 6.2 Particle versus wave experiments

Suppose we remove BS2 completely; then, there is no interference at the BS2 position, and a given photon has a 50–50 chance of being in either of the two arms of the MZI. So, after collecting many photons, we would find half in D1, and half in D2. The latter measurement is analogous to determining which slit the particle went through in the two-slit experiment. So, with BS2 in place, we get wave behavior with only D1 getting counts; without BS2, we get particle behavior with equal counts in D1 and D2. But notice the difference: the wave behavior says that the photon is in both arms simultaneously; the particle behavior says that it is in only one arm at each instance. Which turns out to be the actual event depends on which experiment is done.

To consider further photon experiments, this is a good time to introduce photon polarization, which we will need in Chaps. 10, 12, and 16 as well. Light involves the oscillation of electric and magnetic fields perpendicular to the direction of the light ray. So, the direction of the field is the polarization direction of the light. If the light is traveling horizontally, the polarization might be vertical or horizontal, both perpendicular to the direction of travel. Or it might be at an angle, as shown in Fig. 6.2(*Left*), but still perpendicular to the direction of travel.

(Polarization can also be considered in terms of the spin of the photon, but the classical discussion we give here is easier.) Indeed, the wave function describing a photon with 45° polarization can be considered a superposition of horizontal (H) and vertical (V) polarizations:

$$\psi_{45°} = \frac{1}{\sqrt{2}}(\psi_H + \psi_V). \quad (6.2)$$

A polarizer is a piece of plastic formed with long aligned molecules that passes light with one polarization (perpendicular to the alignment) and absorbs light with the other polarization. Thus, we have something that looks like Fig. 6.2(*Right*). In the figure, we could have the situation where the light is made up of a mixture of photons, some of which have H polarization, and some V; only the H get through the polarizer. Or we could have single photons, each of which is in the state of Eq. (6.2); in that case, each photon has a 50–50 chance of being absorbed or transmitted through. The Malus law states that the probability of getting through depends on the angle the polarization makes with the horizontal (it depends on the square of the cosine of the angle, so it would be unity at 0° and 180°). The photon that gets through will be H polarized.

To build an interferometer, we need mirrors and beam splitters, but the beam splitter we are now going to use is of a special kind: a polarizing beam splitter (PBS), shown in Fig. 6.3. A photon that enters with H polarization will certainly be reflected, and one with V will be transmitted. A photon with a 45° polarization will have a 50 % chance of either case if we measure which path it takes. However, if we do not measure which path, the photon will be in a superposition of the two possibilities. (Note how the final state depends on the measurement process!)

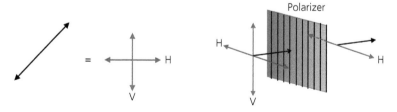

**Figure 6.2** *Left*: Polarization at 45° is a combination of horizontal (H) and vertical (V) polarization. The light is traveling into the page, so the polarization directions are all perpendicular to the direction of travel of the light. *Right*: A polarizer passes only one direction of polarization, which is horizontal in the case shown.

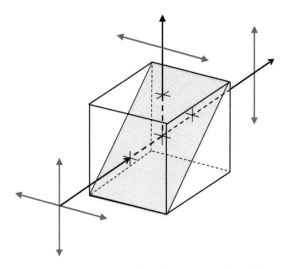

**Figure 6.3** A polarizing beam splitter (PBS) transmits light with vertical polarization and reflects light with horizontal polarization.

A polarization-dependent MZI is shown in Fig. 6.4, where we show the different polarizations of the beams in the two arms. After PBS1, the upper arm carries the H component, and the lower arm carries the V component. If the beam splitter PBS2 is removed, then one

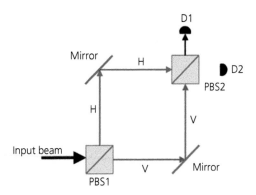

**Figure 6.4** A Mach–Zehnder interferometer constructed with the polarizing beam splitters PBS1 and PBS2. The incoming photons are polarized at 45° and designated by the black arrow. The arms carry either vertically polarized photons (V; indicated in green) or horizontally polarized photons (H; indicated in red).

will have detection in either D1 or D2, depending on which arm the incoming photon takes in this particle-type experiment. However, if the PBS2 is present, then the V arm component is transmitted while the H is reflected, so they are again superposed, and a 45°-polarized photon is detected in D1. Nothing is detected in D2. The case with the PBS2 in place is a wave-type experiment. Although it is the PBS2 that keeps particles out of D2 in this case, and not destructive interference, the final result of a 45°-polarized photon in D1 requires both H and V components to re-superpose. And this will work with one photon at a time in the device.

## 6.3 Delayed choice

In the wave/particle experiments described above, is it possible that the photon somehow "knows" ahead of time, before entering the MZI, whether BS2 is in place and then behaves accordingly? John Wheeler asked this question and suggested a way to get an answer. The idea is to let the photon enter the MZI *before* one decides whether to remove BS2. One can arrange a procedure of quickly removing or leaving in a beam splitter. The result is that it makes no difference; the photon still behaves properly (according to quantum predictions) as a wave or a particle. Again, as Bohr said, it is the actual experiment done that specifies what properties a particle has, that is, *it is the experiment that actualizes the reality.*

To make this delayed choice experiment even more remarkable, we add another feature. It is possible to produce photons in pairs by a process called "spontaneous parametric down-conversion." What occurs is a pair of photons in an entangled state of both in V polarization plus both in H polarization:

$$\psi_{pair} = \frac{1}{\sqrt{2}} \left[ \psi_V(A)\psi_V(B) + \psi_H(A)\psi_H(B) \right], \qquad (6.3)$$

where A refers to a photon emitted to the right going into the MZI, and B refers to the one going left to have its polarization determined. The entire apparatus is shown in Fig. 6.5. Another new feature (besides the two-photon emitter) is the polarization-dependent beam splitter (PDBS) feeding into D1 and D2. This new beam splitter reflects any H photon but splits V photons 50–50, as would a regular beam splitter.

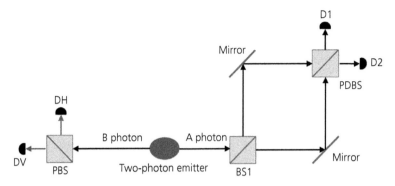

**Figure 6.5** A Mach–Zehnder interferometer with a double-photon emitter as the source. PDBS is a polarization-dependent beam splitter. The entire apparatus is put into an entanglement of particle and wave states; BS1, beam splitter 1; D1, detector 1; D2, detector 2; DH, horizontal detector; DV, vertical detector; PBS, polarizing beam splitter.

Suppose A is an H photon that enters the MZI; it is split equally in a superposition of upper and lower arms, as usual, by BS1; but if it goes in the upper interferometer arm, it is reflected into D1, and if it is in the lower arm it is reflected into D2. Since it is always reflected at PDBS, there is no interference, and we have a *particle* experiment—we can detect which path the photon followed. On the other hand, if a V photon enters the MZI, the components in the two arms interfere at the PDBS; we can arrange for constructive interference for D1, and destructive for D2—this is a *wave* experiment that always sees only D1 detections.

We can arrange that the B photon travels a longer path than the A and does not get to its detector until after the A photon has been detected. Detecting the A photon in D1 does not really tell us its polarization because both the wave and the particle state can have a D1 detection. So, until the B photon is detected, we will not know which kind of experiment we did. We are in an entangled state of wave experiment and particle experiment! If we denote the MZI apparatus's final state by either $\Psi_{\text{wave}}(\text{MZI})$ or $\Psi_{\text{particle}}(\text{MZI})$, then the entangled state is

$$\psi_{\text{total}} = \frac{1}{\sqrt{2}} \left[ \Psi_{\text{wave}}(\text{MZI})\psi_V(B) + \Psi_{\text{particle}}(\text{MZI})\psi_H(B) \right]. \quad (6.4)$$

The wave function collapses to either a particle experiment or a wave experiment only upon the detection of the B photon. This seems to be a macroscopic object in an entangled wave function, rather analogous to the famous Schödinger cat, which we describe in Chap. 8. This interesting experiment was performed in France in 2012 (by F. Kaiser et al) and was even more remarkable by allowing the possible final results to be partial wave/particle states rather than completely one or the other. While particle and wave are complementary ideas, apparently we can have intermediate behaviors between the two.

# 7
# Bell's Theorem and the Mermin Machine

> Bell's theorem is the most profound discovery of science.
>
> HENRY P. STAPP

Bell's theorem is one of the most remarkable proofs in physics, since it shows that a whole class of theories, those that are local and realistic, must obey restrictions on certain of their experimental predictions that quantum mechanical theories violate. This violation, when verified experimentally, shows that nature must be either nonrealistic or nonlocal (having "action" at a distance). Basically the result is that, while the realism of a hidden-variables theory might still be possible, nonlocality would then be needed—as it is in the Bohm hidden-variables theory, discussed in Sec. 5.4.2. Most physicists accept nonlocality and have no particular need to add hidden-variables realism to a quantum theory interpretation, since such as addition seems to give no different experimental predictions.

## 7.1 The Mermin machine

A hypothetical machine invented by David Mermin is meant to demonstrate how a local, realistic interpretation of a relatively simple experiment gets a result that is at odds with quantum mechanics. We will set up the experiment and discuss a classical hidden-variable interpretation and show that it contradicts the quantum prediction. In the experiment, Alice and Bob each have a box that makes measurements on pairs of particles emitted from a central source. A detector has a switch that allows three settings, 1, 2, or 3; in the experiment, either a red light or a green light flashes, depending on the result of the experiment. See Fig. 7.1.

In real experiments, the box is measuring either the orientation of the spin of a particle or the polarization of a photon. The source is emitting

# What quantum mechanics predicts

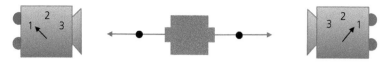

**Figure 7.1** The Mermin device. The green source in the middle emits a pair of particles, one to each detector. Each detector has three settings and lights that flash red or green, depending on the result of the experiment.

entangled pairs of particles in spin singlet states, or pairs of photons with correlated polarizations.

## 7.2 What quantum mechanics predicts

Let's look at the device as if it measured spins. Bob and Alice each have an SGA, as we discussed in Chap. 5. Suppose we have an SGA that is aligned along the $z$-axis. We shoot a particle whose spin is aligned by a previous experiment along an axis called $u$ that is at an angle $\theta$ with $z$, as shown in Fig. 7.2. When this measurement is done, we get either up along $z$ or down along $z$. What quantum mechanics tells us is that the probability of finding an up measurement along $z$ of a incoming spin that is known to be along $u$ is given by

$$P_{\uparrow z}(\uparrow u) = \left(\cos\frac{\theta}{2}\right)^2, \tag{7.1}$$

where we are using the cosine of half the angle and then squaring that result. How quantum mechanics gets this result requires a good deal of theory that we cannot cover here. However, the derivation is very

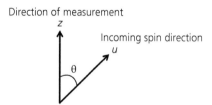

**Figure 7.2** We set up a Stern–Gerlach apparatus (SGA) to measure the spin along $z$. A spin known to be aligned along $u$ enters the apparatus. The SGA gives a result that is either up along $z$ or down along $z$.

logically based on general quantum principles. Suppose our measurement angle is $\theta = 60°$; then, the cosine of $30°$ is $\sqrt{3}/2$, so the probability of finding the spin up is $3/4$. What this says is that three out of every four measurements will on average show an up result. On the other hand, the probability of finding a down measurement along $z$ of a spin that comes in aligned along $u$ is given by

$$P_{\downarrow z}(\uparrow u) = \left(\sin \frac{\theta}{2}\right)^2. \tag{7.2}$$

Here, we use the sine function. Then, since $\sin 30° = 1/2$, the probability of finding the spin down along $z$ is $1/4$. We will find one out of four down spins in our measurement on average. Note that the two probabilities add to unity: $3/4 + 1/4 = 4/4 = 1$. This is because all the particles are either up or down—the sum of the probabilities of all the events is always one, that is, four out of every four.

When we interpret the Mermin machine as a pair of Stern–Gerlach devices, the settings 1, 2, and 3 will correspond to angles $0°$, $120°$, and $240°$, respectively, as shown in Fig. 7.3. Because the spins in the singlet spin state are always opposite to each other, we coordinate Alice's settings and Bob's settings oppositely, as shown in the figure. Now, if both boxes are put on setting 1, and a red light occurs in box A, then we

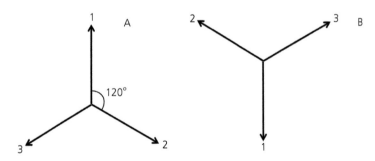

**Figure 7.3** The settings 1, 2, and 3 on the Mermin machine boxes A and B measure spins along the three directions shown, with each pair of directions separated by $120°$ angles, as shown. If a box is put on setting 2, it measures whether an incoming spin is up (so that a red light flashes) or down (so that a green light flashes) along direction 2. Because the two spins in our singlet state have one up and one down spin along any direction, the measurement of up along 1 in box A will correspond to up along the oppositely directed 1 in B.

know the spin is up along direction 1, and we expect a red light in box B as well.

The quantum probabilities for a *single* particle entering an SGA (say box A) with spin up along $z$ (which we take as direction A1) are, for the three possible device settings, given by Eqs. (7.1) and (7.2) as follows:

$$P_{\uparrow z}(\uparrow 0°) = 1, \tag{7.3}$$

$$P_{\uparrow z}(\uparrow 120°) = \frac{1}{4}, \tag{7.4}$$

$$P_{\uparrow z}(\uparrow 240°) = \frac{1}{4}, \tag{7.5}$$

$$P_{\uparrow z}(\downarrow 0°) = 0, \tag{7.6}$$

$$P_{\uparrow z}(\downarrow 120°) = \frac{3}{4}, \text{ and} \tag{7.7}$$

$$P_{\uparrow z}(\downarrow 240°) = \frac{3}{4}. \tag{7.8}$$

Let's next compute the probabilities that quantum mechanics gives for the whole Mermin device. We can always describe our incoming wave function by the one in Eq. (5.8) in Sec. 5.4, as this one says that the first particle is up along $z$ (we take that as direction A1), with the second down along $z$, OR that the first is down along $z$, and the other is up along $z$:

$$\psi_{\text{singlet}} = \frac{1}{\sqrt{2}} \left( e_{\uparrow z} p_{\downarrow z} - e_{\downarrow z} p_{\uparrow z} \right). \tag{7.9}$$

Recall that singlet rotational symmetry allows us to replace $z$ in this wave function by any direction $u$.

So, if the setting is 1 on both boxes and we get a red light on A, then we will certainly get a red light on B. The spin angle $\theta = 0$, in this case, and $\cos 0 = 1$. But it is equally likely that we will get a green light on A (down spin) because of the second term in Eq. (7.9), but we will also get green on B then. We never get a red and a green together in this case. Thus, in the case (call it type *a*) where *both setting are the same,* here both on 1, we will certainly have

$$P(RR) = P(GG) = \frac{1}{2}, \text{ and} \tag{7.10}$$
$$P(RG) = P(GR) = 0.$$

The same result will occur if both box settings are 2 or if they are both 3. If box A gives an up (red) result, we know box B must also give a red result, given its opposite spin arrangement. So, we will get RR or GG each time here too, and the same probabilities result.

Now, suppose we have box A set at 1, and box B at 2, as illustrated in Fig. 7.4. Call such *unequal* settings type b. The probability of getting an R on A is 1/2 (equally likely up or down), but the probability of getting R on B is given by Eq. (7.4) as 1/4, so the probability of RR is (1/2)(1/4) = 1/8. We will get this result (A1 up, B2 up—i.e., RR) one out of any eight tries on average. If we ask for the probability of getting RG in this case, we need the probability of getting down for B along 2; this probability is 3/4 (Eq. [7.7]), so the joint probability is 1/2 × 3/4 = 3/8. The same arguments go through if we ask for the likelihood of getting

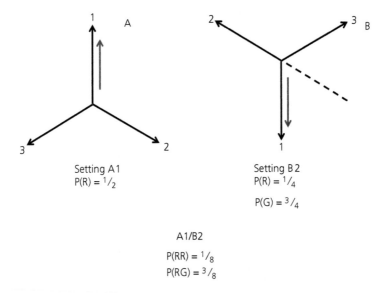

Setting A1
P(R) = 1/2

Setting B2
P(R) = 1/4
P(G) = 3/4

A1/B2
P(RR) = 1/8
P(RG) = 3/8

**Figure 7.4** We consider an up/down spin pair with the 1 setting on box A, and the 2 setting on box B. The probability of getting A up along 1 (R; as shown) is 1/2, and that of setting B as up along 2 is 1/4, so the joint probability is the product, or 1/8. The probability of B being down along 2, that is, along the dotted line, is 3/4, so the joint probability is 3/8.

G (down) on the A1 device. So, we have type b with *different* settings of the two boxes:

$$P(RR) = P(GG) = \frac{1}{8}, \text{ and} \tag{7.11}$$
$$P(RG) = P(GR) = \frac{3}{8}.$$

Adding the four probabilities together gives 1, as it should. We did all the calculations for device A set on 1, but everything goes through identically if we use A2 or A3 because we can take a singlet state's up direction along any axis.

The main result of this section is that the quantum prediction is that, in type b experiments, the same color (RR or GG) occurs in $1/8 + 1/8 = 1/4$ of the experiments.

## 7.3 What a hidden-variables theory might say

But now let's look at an explicit hidden-variable calculation of the same thing. We suppose that each particle carries with it some property (a code "written" on the particle) that determines what light will flash at each setting. The code will be different each time a pair of particles is

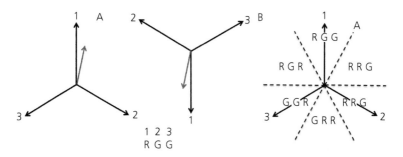

**Figure 7.5** We consider an up/down spin pair as seen by Alice and Bob. The pair has imprinted on it the hidden-variable codes for any possible settings Alice and Bob choose in the local realistic model. If Alice's spin is in the angular region around direction 1, she will get R on setting 1, and G if she chooses settings 2 or 3. The same codes will be on Bob's particle, giving the light color corresponding to his setting.

sent out, according to how the two particles are correlated at the source, but, after that, they are on their own. On a single particle, the possible imprinted codes telling what light will flash, given switch setting 1, 2, or 3, are RRR, RRG, RGR, GRR, RGG, GRG, GGR, and GGG. Fig. 7.5 shows how these choices might come about.

The two spins are again correlated to be opposite but could be along any direction, as shown. The first figure shows spin A at a small angle with the vertical. Its components (whether it is more along or opposite to the direction) along the three axes at 120° to one another are positive, negative, and negative, respectively, so it would register RGG on the lights in the three switch settings. The second figure shows particle B to have exactly opposite spin, so its components are also positive, negative, and negative, and the B device is set to register RGG there too. The third figure gives all the R or G arrangements when a spin falls within the various 60° segments around the full 360°. Note that RRR and GGG *never* occur in this particular hidden-variable approach.

What correlation must the particles have at the source to explain data of type a where the switch settings are the same? Apparently, the two particles start out with identical codes. If particle A has RGR, then particle B has RGR. Otherwise, how could we ensure that we always got an RR or a GG and never RG or GR? Such an arrangement is what we get with the hidden-variable model.

But the above explanation of type a data has implications for type b data. Here are the codes for the two particles:

A: RRG, RGR, GRR, RGG, GRG, GGR
B: RRG, RGR, GRR, RGG, GRG, GGR

(We have eliminated RRR and GGG for the time being, since they were not possible in our hidden-variable approach. We will later see what effect they would have had upon inclusion.) Suppose the switch setting is 12, that is, 1 for A, and 2 for B; count the number of RR, RG, GR, and GG flashes for this switch arrangement. (Draw a line between the first letter in each item of the A list and the second letter in the corresponding item in the B list.) The first combination (RRG) gives RR for this setting (data output [12RR]), the second gives (12RG), the third (12RG), etc. Counting up, you get the results shown in Table 7.1.

The same count occurs for each of the other possible settings of type b, such as 21, 23, etc. Thus, one sees that the same color flashed 2/6 = 1/3 of the time. (If RRR and GGG are allowed, they contribute another

**Table 7.1** Relative numbers of flashes when the switch setting is 12

| RR | RG | GR | GG |
|----|----|----|----|
| 1  | 2  | 2  | 1  |

RR and GG, so that the same color flashes 1/2 of the time). Note that allowing these extra possibilities strengthens the argument we are giving, but eliminating any more of the possibilities would weaken it. So we can say that the same color flashes *at least 1/3 of the time*, with built-in instruction codes on the particles in type b data. However, quantum mechanics (and experiment) gives, for this data set, $P(RR) + P(GG) = 1/8 + 1/8 = 1/4$, which is less than the 1/3 (or 1/2) of the "common sense" view. The local-realistic assumption of the built-in instructions sets predicts *too many* of the same color flashes. It is in contradiction with the quantum result, which we will see is in agreement with experiment.

This discussion of the Mermin device is not a rigorous proof of the impossibility of devising a hidden-variable theory that is based on local realism and that agrees with the quantum prediction. That discussion was given by John Bell in his famous set of inequalities, which established limits on experiments. However, precise experiments in recent years have shown rather convincingly that nature violates those inequalities, demonstrating the impossibility of hidden-variable theories that are both local and realistic. We study Bell's inequality more rigorously in Sec. 7.4.

## 7.4 A Bell inequality

Let's consider a slightly different set up of the measurement devices. Suppose Alice uses either a setting $a$ at angle $0°$ or a setting $a'$ at $120°$. Similarly, Bob uses $b = 0°$, and $b' = 120°$. When Alice gets a red light (up spin) on setting $a$, she writes down $A = +1$ and when she gets a green light she writes down $A = -1$; similarly, for setting $a'$, she writes $A'$ as equal to either $+1$ or $-1$, depending on the lights. Bob also records $B$ and $B'$ as $+1$ or $-1$. Alice and Bob can each use either of the two settings—and decide at the last instant which to use. The particles coming to their detectors, if there are hidden variables, must be "prepared" at the source for each of the four possible pairs of settings and so that we can consider what the products of their numbers $A, A', B, B'$ might be. We form

$$AB + AB' + A'B - A'B' = A(B + B') + A'(B - B'). \qquad (7.12)$$

The values of each of the individual numbers are either $+1$ or $-1$, so either $B + B'$ or $B - B'$ must vanish (e.g., if $B = 1$, and $B' = -1$, then $B + B' = 0$, and $B - B' = 2$). Since the factor $A$ or $A'$ multiplying $B + B'$, or $B - B'$, is $\pm 1$, we must have the quantity lying between $-2$ and $+2$:

$$-2 \leq AB + AB' + A'B - A'B' \leq 2. \tag{7.13}$$

What we now do is to collect a lot of data and take averages over many experiments. If $\langle AB \rangle$ is the average of the product when both angles are at zero, then we must have the experimental result

$$-2 \leq \langle AB \rangle + \langle AB' \rangle + \langle A'B \rangle - \langle A'B' \rangle \leq 2. \tag{7.14}$$

This particular expression for testing hidden variables was suggested by John Clauser, Michael Horne, Abner Shimony, and Richard Holt and is called the CHSH inequality. John Bell originally suggested the idea of a test of local reality by an inequality, although different from the one shown here. An experimental check of Eq. (7.14), with angles set to make the quantity in the middle as large as possible, found it to be 2.697, which clearly violates the inequality. So, how can quantum mechanics violate such a simple mathematical rule? The answer is that quantum mechanics involves only one experiment at a time. The pair of spins in a hidden-variable analysis must be ready for *any* experimental setting that could be switched to at the last instant; so every experiment is included in the hidden variables of each firing of the source. But, in quantum mechanics, *unperformed experiments have no results*. We either perform the $AB$-type experiment, an $A'B$-type experiment, or one of the others, and not all at the same time. Each experiment is independent of the others, and so we can't really say that either $B + B'$ or $B - B'$ must vanish. The correlations are instantaneous, depending on the set up Alice and Bob choose, and the averages depend only on the particular set up that presently exists at each reading. The local realistic approach of hidden variables is disproved by experiment. Quantum mechanics is beautifully successful in predicting the numerical result that the experiment actually gives, although we have not shown the details of the calculation here.

In summary, then, the Bell theorem assumes that certain "elements of reality" exist, that is, the hidden variables. It also assumes that interactions are local, that is, a setting of the hidden variables takes place when the particles are created together and once they separate, there can no longer be any adjustment of the parameters. These assumptions lead to predictions that are violated by quantum predictions and by experiment, so that the assumptions are invalid.

The experiments were originally done by Stuart Freedman and John Clauser in 1972, and then by Edward Fry and Randall Thompson in 1976 and later, more precisely, by Alain Aspect, Phillipe Grangier, and Gérard Roger in 1981 and 1982. Others have followed. All of the experiments have had loopholes, that is, one can think of logical possibilities, even if remote, that an alternative explanation besides that given by quantum mechanics might still be possible. One loophole is having the detectors close enough that there could have been some sort of communication between them, correlating the results to give the quantum results while avoiding the need for an instantaneous collapse of the wave function upon the first of the measurements. Recently, this loophole has been eliminated by an experiment by researchers from Delft University of Technology. The detectors in this experiment were separated by 1.3 km, so that any correlations at the speed of light were impossible. Einstein decried the "spooky action at a distance," but apparently it is there.

#  8

## What Is a Wave Function?

> Man has to postulate weirdness, before reaching the new science.
>
> TOBA BETA

We have discussed wave functions and their connection with the probability of, say, finding a particle at a location or in a certain spin state. Wave functions are always connected with probability, but what exactly are they physically? Is the wave function a physical quantity like mass and energy, or is it just something in the mind of the physicist, measuring his state of knowledge about the particle? The distinction is made between the "ontic" interpretations and the "epistemic" interpretations. The ontic theories consider the wave function as describing a real physical quantity or, rather, as being itself a real physical quantity, while the epistemic theories, consider the wave function as being a measure of human belief about a quantum system. There is considerable current debate on this distinction; we need to look into the issue more closely.

## 8.1 Schrödinger's cat

A good place to start is with Schrödinger's cat. We discussed superposition in Chap. 4, and entanglement in Chap. 5. When we talked about a particle passing through a double slit or two electrons in a singlet state, these were pretty amazing ideas, and they have been verified experimentally. But can we use these same ideas to describe large objects? Bohr, in his interpretation of quantum mechanics (known as the Copenhagen interpretation), considered the world to be divided into the microscopic quantum world and the classical large-scale world. When a measurement is made, the two worlds interact, and a large measuring apparatus with some kind of attached screen or meter records what happened. For example, in the Stern–Gerlach experiment, a spin passes through a large magnet, gets deflected up or down depending on the direction of the spin, and hits a screen where a bright spot occurs, telling us whether the spin was up or down. We associate a position on the screen with the microscopic direction of the spin. The magnet and screen can be described by classical physics, while quantum mechanics is needed to describe the spin.

But where is the dividing line here between big and small? Can we use a wave function to describe our measuring device? Why does quantum mechanics not describe big objects (even humans)—or does it? In fact, experiments are going on now to see how large an object can be and still be made to show quantum effects. The results are finding larger and larger systems that demonstrate effects like quantum interference. A Bose–Einstein condensate (discussed in Chap. 9) is a collection of hundreds of thousands to millions of atoms in a clump about 10–50 micrometers across (near the diameter of a thin human hair); in many ways, it acts something like a large molecule, requiring quantum mechanics to describe it. It shows all the wave characteristics of quantum systems, but to do so it must be at remarkably low temperatures—much lower than any other place in the universe.

Schrödinger gave an example of the application of quantum mechanics to a large-scale system, his cat, to demonstrate that one could seemingly get into a conceptual problem by blindly applying quantum mechanics to large systems. In this thought experiment, we put his cat into a box, but we also place a radioactive atom, a Geiger counter, and a bottle of cyanide into the same box. The radioactive atom, when it decays, triggers an electrical pulse in the counter, which opens a valve to release the poison gas, killing the poor cat.

If we can describe the cat by a single pure quantum state (a questionable assumption), then the initial state is cat alive ($C_A$) and atom undecayed ($a_u$):

$$\Psi_{\text{initial}} = C_A a_u. \tag{8.1}$$

Suppose the half-life of the atom is such that, after 1 hour, the probability is 1/2 that the atom has decayed, and 1/2 that is still undecayed. Then after that 1 hour the state must be

$$\Psi_{1\text{ hour}} = \frac{1}{\sqrt{2}}(C_A a_u + C_D a_d), \tag{8.2}$$

where $C_D a_d$ represents a dead cat and the decayed atom. The system probability function, which is

$$(\Psi_{1\text{ hour}})^2 = \frac{1}{2}(C_A a_u)^2 + \frac{1}{2}(C_D a_d)^2 + C_A a_u C_D a_d, \qquad (8.3)$$

gives the probability of finding the cat dead as 1/2, and the probability of finding it alive as 1/2, but also has *interference between aliveness and deadness* in the last term. In actual practice, were we to open the box, we are pretty certain we would never observe such a peculiar blurred superposition of alive and dead states as given in Eq. (8.2). In the particular experiment described, the interference term should disappear because the atomic wave functions $a_u$ and $a_d$ are so different that their overlap will vanish, but we don't *expect* to see any interference terms in experiments involving macroscopic objects. Why does the principle that seems to apply to the microscopic world and gives it its weirdness, not apply to the macroscopic world of cats? As the science writer David Lindley asks, "Where does the weirdness go?"

When we open the box, we see either an alive cat or a dead cat. The act of observing seems to have *collapsed* the wave function to one of the two possible states or the other. We have seen wave function collapse before for microscopic systems; it is a postulate of quantum mechanics that a superposition does collapse to one of the member states upon measurement, but it remains mysterious. What is it that constitutes the act of observing? Is it the act of opening the box? Suppose I open the box, but don't actually look to see the state of the cat. Is it my act of looking that collapses the wave function? Is it my knowledge of the state that does it?

A doctor comes into waiting room of a veterinary hospital to tell Schrödinger the condition of his cat. Can you imagine how she might present the strange medical condition after the experiment was done? The good news about the philosophic problems that are faced in quantum mechanics is that they have stimulated considerable recent research. The differences between wave functions of microscopic and macroscopic objects have been analyzed recently with some success via the idea of decoherence, which we discuss in Sec. 8.2. But the bad news is that many questions involved in the measurement process have not been solved to everyone's (anyone's?) satisfaction.

The question of what constitutes an observation is a difficult problem as was illustrated by Eugene Wigner.

## 8.2 Wigner's friend

Suppose the famous physicist Eugene Wigner has a friend, Grace, who is the observer who opens the cat box. Before opening it, she would say the wave function is given by Eq. (8.2), an entangled state of the atom and cat. When she opens the box, she finds the cat is alive, and the wave function collapses to $C_A a_u$ by her measurement. (If describing a cat by a wave function bothers you, as it might, just consider the atom by itself in the box in a superposition of undecayed and decayed.) Meanwhile, Wigner has remained outside the room and does not know the result of the box being opened. He might claim the wave function is given by

$$\Psi_{1\text{ hour}} = \frac{1}{\sqrt{2}}(G_H C_A a_u + G_S C_D a_d), \tag{8.4}$$

where $G_H$ and $G_S$ are the quantum states of Grace, happy or sad, according to the state of the alive cat or the dead cat, respectively. Wigner enters the room and finds that Grace is happy and the cat is still alive. Now Wigner might say that the wave function collapsed only when he entered the room. Of course, *another* friend of Wigner's who comes into the lab even a bit later could say that the collapse happened when he observed Wigner observing Grace. There could be further observers, in an infinite regress. What about the wave function of the entire universe? Who collapses that?!? The philosophic problem of where wave function collapse actually happens is a difficult one.

The case of Wigner's friend gives rise to an old joke in physics: Scientist A is doing an experiment that has a good chance of finding a result that would win a Nobel Prize. But, again, maybe it won't work. He works hard for months and finally gets the great result. Scientist B, a theorist, who has had little to do with the project walks into the lab and A tells him excitedly of the result. B says, "Yes I agree this result is wonderful and that it may be worth a Nobel Prize, and I want my name on the paper announcing the result, because it was I who collapsed your wave function and that of the experiment from a superposition state of discovery and non discovery into the successful state by coming into your lab." Of course, no experimentalist would ever accept this argument.

The Copenhagen interpretation has answers to some of the questions we have raised above by dividing the universe into microscopic and macroscopic. There are alternative theories and interpretations as well. Among these are the statistical interpretation with decoherence,

Bohm's interpretation, Everett's many-universes view, and the nonlinear Schrödinger equation method.

The Everett interpretation is perhaps the most bizarre. In this view, the evolution of a wave function superposition upon measurement results in each possibility occurring in a separate universe. In one of them, the cat is dead, and I witness that result; in the other, the cat remains alive and another "me" witnesses *that* result. The nice feature of this idea is that the mathematics is consistent; the wave function evolves by the Schrodinger equation normally, and one does not have to add a new feature, namely, the collapse of the wave function upon measurement in our single universe. There are many serious supporters of this unusual approach. Unfortunately, since the universes never interact, there seems no way to experimentally test this interpretation. Also, given that every possibility actually occurs, one wonders where the probability rule of quantum mechanics fits in.

Recently, the *Bayesian* probability approach to interpreting the quantum mechanical wave function has gained a following. In this epistemic view, the wave function represents an experimenter's best bet of what the state of matter is. (It is actually based on probability ideas involving betting.) Thus, the wave function is a property of the experimenter's mind and not a physical reality. Most physicists have not yet accepted this approach; the wave function behaves too much like it is really out there. We will examine this approach in more detail in Sec. 8.5.

## 8.3 Quantum interference of macroscopically distinct objects

Nobel Prize winner Tony Leggett has emphasized that an essential goal of experiments should be to show the quantumness of increasingly large objects by showing they can be involved in quantum interference—much like the interference pattern that develops when a photon or atom goes through the double slit. Several experiments are seeking to do just this. For example, carbon-60 atoms ("buckyballs") have been used in the double-slit experiment instead of electrons, and an interference pattern has been seen. Even heavier organic molecules have been used, with success. The most impressive of these experiments has involved the use of a SQUID, which stands for "superconducting quantum interference device." In this device, a metal loop is taken to a very low temperature, at which point the electrons in the metal become superconducting—able to conduct electricity without resistance. A small break in the ring requires the electrons to quantum tunnel across the barrier but also allows the establishment of what is called a phase difference across the barrier. The ring can trap a quantized amount of magnetic field within the loop, with this corresponding to a set amount of current flowing clockwise through the loop. Of course, the current could equally well be going counterclockwise, with the magnetic field penetrating the opposite direction. Being a good quantum system, the loop can be suspended in a superposition of the two rotating states simultaneously, with one state corresponding to a unit of current going one way, and the other state corresponding to the opposite flow. One can observe the interference between these two states. But this is not just one electron in two states but involves more than a million million electrons, all behaving coherently like one quantum system. It is not quite a whole cat, but it is a very large object.

The interferometer shown in Fig. 6.5 might seem to qualify as being in a cat state, although we did not see any interference between the two parts of the wave and particle entangled states. Serge Haroche, who won the Nobel Prize in 2012, has created Schrödinger "cats" of photons, with superposition of small numbers of photons in two different "coherent states" showing interference. There have been proposals to see interfering "cats" made up of a million atoms in a Bose–Einstein condensate, which might readily show the interference. Again, the atoms are in an extremely low temperature state where they are able to maintain quantum coherence and avoid the outside influences that could destroy this coherence. We turn to the question of such decoherence now.

## 8.4 Decoherence

In an example like that of the Schrodinger's cat experiment, in which a macroscopic object is one of the partners in an entangled state, one can ask, as we did in Sec. 8.3, why it is so difficult to see such a superposition of states, that is, simultaneously see the live and dead cat. The wave function seems to have collapsed to a single term by the time an experimenter looks at the system. A related, but slightly different, question is why we never see the interference between the various members of a superposition of macroscopic objects. Interference requires a certain quantum *coherence* to be maintained in an experiment, just as classical light sources have to be coherent to show interference in a diffraction experiment. When an object becomes large, it becomes difficult to isolate it from the outside world, and those interactions destroy the coherence, thus removing any interference effects from the probability distribution. Therefore, we would not expect to see interference between a live cat state and a dead cat state. We need to discuss these two distinct philosophical problems: removal of the interference, and the collapse of the wave function.

Consider again the two-slit problem. Recall that the probability of an electron hitting a screen after passing through the double slit was

$$P_{\text{electron}} = \left[\frac{1}{\sqrt{2}}(\psi_1 + \psi_2)\right]^2 = \frac{1}{2}\left(\psi_1^2 + \psi_2^2 + 2\psi_1\psi_2\right). \quad (8.5)$$

The first two terms represent the probability of going through slit 1 and slit 2 (both terms are positive), but the last term can be positive or negative, depending on position, so that it causes the probability and the resulting display on a screen to show brightness oscillations that represent the interference. We want to describe how interference is removed when we try to determine through which slit the electron passed by means of shining light on the electron as it passes through the slits. Fig. 8.1 shows a possible arrangement. The light hits the electron, and if the latter went through slit 1 or slit 2, the scattered photon follows a trajectory that we can trace back to its origin near one or the other slit. We might have a camera to detect the position of the photon. When we look at the image, we will be able to see through which slit the electron passed. We describe this distribution of originating positions by a photon wave function $p_i$ ($i = 1, 2$), with the variable $x_i$ describing positions on the camera image that tell us that the electron was in the vicinity of slit $i$. Before we observe the photon, the electron and photon are then entangled according to

$$\Psi_{\text{electron/photon}} = \frac{1}{\sqrt{2}}\left(\psi_1 p_1 + \psi_2 p_2\right). \quad (8.6)$$

If the function is peaked up around a position at $x_1$, as shown by the camera image, it indicates that the electron would have passed in the vicinity of slit 1, and similarly for $x_2$. Before we detect the photon, the probability function for combined electron and photon is

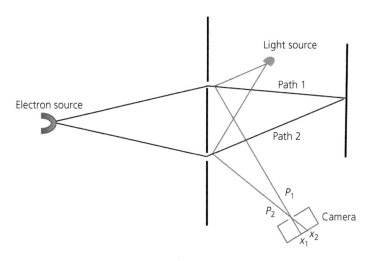

**Figure 8.1** An arrangement with a light source and a camera to detect through which slit an electron passes. If the electron follows path 1, then the scattered photon ends at $x_1$ in the camera with wave function $p_1$; if path 2 is followed, the photon is scattered into $p_2$.

$$P_{el/ph} = \frac{1}{2}\left(\psi_1 p_1 + \psi_2 p_2\right)^2 = \frac{1}{2}\left(\psi_1^2 p_1^2 + \psi_2^2 p_2^2 + 2\psi_1\psi_2 p_2 p_1\right). \quad (8.7)$$

The cross term in this will give us interference unless it vanishes. However, suppose our camera system has not yet detected the photon. We *could* have looked at it, so it is *knowable*, but we have not yet placed the camera to catch the photon. We will merely observe the electron's position when it hits its screen. A typical plot of $p_1$ and $p_2$ is shown in Fig. 8.2. Under the conditions shown (we used a photon of short-enough wavelength), we certainly would have been able to tell which slit the electron went through if we had examined the camera image. That is, $p_1$ and $p_2$ are each localized around their respective positions, and the two functions do not overlap. The result is that the product $p_1 p_2$ vanishes and, with it, the interference term. Our *potential* knowledge of the electron slit has ruined the interference, even though we did not actually use that information to determine the slit. All we get on the electron's screen (after many such electrons) is the overlapping distributions of electrons from each of the two slits, $\psi_i^2$. Such a situation has been termed "information leakage"; the photon has taken something vital away with it, *ruining the electron coherence* and the interference, even

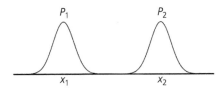

**Figure 8.2** A plot of the wave functions for a photon, showing the probability of the photon appearing at $x$ on a camera image when we use a short-wavelength photon. If it appears in the neighborhood of $x_1$ on the image, then we know the electron went through slit 1; if it appears near $x_2$, the electron went through slit 2. The two positions are well separated, so we can definitely distinguish the slit.

when we did not detect the photon. In the same way, most systems are in constant interaction with the rest of the universe and suffer such *decoherence*, thus losing some of their quantum effects. The larger the object, the more it interacts with the outside world, and the faster it decoheres. Maintaining the quantum character of a large object is very difficult; for example, our universe is filled with radiation—even heat radiation, for example—constantly bouncing off things. So, isolating an object requires great care and often low temperatures.

Other scenarios are possible. Suppose the photon had a long wavelength, so that examination of it would not have revealed with certainty through which slit the electron passed. This would be the case if the $p_i$ functions were each very broad and overlapped considerably, as shown in Fig. 8.3. Depending on how large the overlap function $p_2 p_1$ is, we may get some or most of the interference term remaining in Eq. (8.7). If one does not get perfect slit information, then one does not totally remove the interference pattern. There is complementarity between

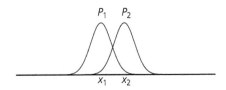

**Figure 8.3** A plot of the wave functions for a photon showing the probability of the photon appearing at $x$ on a camera image when we use a photon wavelength that is too large. The wave functions overlap, so it will be difficult to distinguish the slit through which the electron passed.

wave nature (seeing the interference pattern) and particle nature (seeing which slit the particle passed through), but sometimes it is a matter of degree rather than all or nothing.

There is another way we may lose interference terms: by lack of coherence of the source. Suppose we send an electron that is more likely to go through one slit than the other. A general wave function for the $i$th electron like this has the form

$$\psi_{\text{electron}} = (\alpha_i \psi_1 + \beta_i \psi_2), \tag{8.8}$$

in which the probability of the $i$th electron going through slit 1 is $\alpha_i^2$, and the probability of it going through slit 2 is $\beta_i^2$. Since, if observed, it went through one or the other, we must have $\alpha_i^2 + \beta_i^2 = 1$. The probability for the $i$th electron is

$$P_i = (\alpha_i \psi_1 + \beta_i \psi_2)^2. \tag{8.9}$$

Suppose, instead of having all our electrons having the same wave function, we have a group of $M$ electrons that have varying $\alpha$ and $\beta$; that is, we have an *incoherent* source. We have to average the probabilities over all the different possibilities so we have the overall probability as

$$P = \frac{1}{M} \sum_i (\alpha_i \psi_1 + \beta_i \psi_2)^2 \tag{8.10}$$

$$= \frac{1}{M} \sum_i \left( \alpha_i^2 \psi_1^2 + \beta_i^2 \psi_2^2 + 2\alpha_i \beta_i \psi_1 \psi_2 \right),$$

where the $\sum_i$ means we are adding up all the various terms. Such a collection is said to be in a *mixed state*. Each electron is in one or the other state $i$, but we do not know which one it will be—rather like the classical case of a thrown pair of dice falling in one of 36 unpredictable number combinations. In a random mixture, we will have $\alpha_i$ and $\beta_i$ as often positive as negative, so that the sum over a large collection of possibilities will have $\frac{1}{M} \sum_i \alpha_i \beta_i = 0$. We also will have $\frac{1}{M} \sum_i \alpha_i^2 = \frac{1}{M} \sum_i \beta_i^2 = \frac{1}{2}$ (it is equally likely for the electron to pass through either slit). The interference terms will average out, and our final result will be just

$$P = \frac{1}{2} \left( \psi_1^2 + \psi_2^2 \right). \tag{8.11}$$

The screen will show two wide bumps corresponding to each of the two slits, with no interference oscillations. It would seem more likely of a large object like a cat to be in a mixed state rather than a pure state described by a unique wave function.

Let's go back to Eq. (8.6) and now view the photon image in the camera so that we actually tell which slit the electron passed through. Then, our result is either $\psi_1 p_1$ or $\psi_2 p_2$, each with probability 1/2. The wave function has collapsed to one of the results or the other. In the Copenhagen interpretation of quantum mechanics, this collapse is a separate assumption of quantum mechanics; it does not happen as a result of the Schrödinger equation.

Let's compare this to what happens in classical probability theory. In this, one is never surprised to find that only one of the many possible occurrences in a probability distribution actually shows up in a trial; there is no need to consider a phenomenon analogous to the collapse of the wave packet. For example, if I want to find the probability that heads or tails will turn up in the flip of a coin, I assemble a large ensemble of coins, throw them, and count the number of heads and tails that show up. Each side is equally likely, but I am not amazed that, in a single throw, a head appears. There has been no collapse of any wave function. The equal-chance probability distribution is a measure of my lack of *knowledge* of the initial conditions on the particular throw—how much force my hand used, what side was up in my hand initially, etc.—and not of a fundamental indeterminacy of the dynamics of the throw of a single coin.

But, with the usual interpretation of quantum probabilities, even with decoherence ruining the interference term, we must still assume a collapse of the wave function to one or another of the terms in the superposition when we do a measurement. This need for collapse to actualize the reality is still a controversial subject in quantum mechanics. Decoherence seems not enough by itself—in contrast to the claims of several of its proponents. There have been several of proposals to solve the collapse problem.

One approach that I personally find appealing was originally suggested by Gian Carlo Ghirardi, Alberto Rimini, and Tullio Weber and further developed by Philip Pearle in a theory (the GRWP theory) called "continuous spontaneous localization." Here, the collapse of the wave function is a *dynamical* process described by an altered Schrödinger equation. A nice thing about such an ontic theory is that the wave function remains a physical object; moreover, the theory can be tested by experiment. Randomly, one or the other of the terms in the superposition physically decays away, leaving only one actualization. The more macroscopic the objects in the superposition, the more rapid the decay, just as we would expect. When one makes a measurement, any microscopic elements become entangled with a macroscopic measuring device, as when the photon detecting the slit enters the camera, so that collapse is almost instantaneous. There are two critical parameters in the theory that, in principle, can be estimated in very sensitive experiments. Presently, all we know is that the parameters are very small, but we cannot yet say that they are zero, as they would be in standard quantum theory. Present experiments give an upper limit on the rate of collapse, that is, they tell us the parameter must be less than some specific value. What one wants is to be able to say that the parameter is *larger than zero*, since it is zero if standard quantum mechanics holds.

Up to now, we have been assuming that the wave function is a real physical object—although a peculiar one; this is the "ontic" view. There is a school of quantum thought that holds that the wave function is nothing more than a *measure of our knowledge* of the state of our experimental system (the "epistemic" approach). One of the versions of such a proposal is based on a mathematical theory called Bayesian probability, and the developers of this theory are "quantum Bayesianists," or QBists; we discuss their ideas next.

## 8.5 Quantum Bayesianism

There are two ways to look at probability: the frequentist way, and that of Thomas Bayes, who lived in the first half of the eighteenth century. In the first way, probability is a fraction: the number of occurrences of a specific event over the total number of possible outcomes. For example, the frequentist would say that, out of 50 coin tosses, the probable number of heads is 25, and the probability of heads is 0.50. The probability of throwing a 2 on a pair of dice is 1/36, because there are 36 ways dice can appear and only one of them is 2. So, probability is a measure of physical events.

On the other hand, Bayes's view is that probability is not a thing external to the person calculating the probability but rather is a measure of that person's *belief* in what will happen. It might be a measure of the faith a gambler has of winning money when he places a bet. To show what this means, let's imagine a conversation between F, a frequentist, and B, a Bayesian.

F: If I flip an infinite number of fairly weighted coins half of them will come up heads.

B: But you can't flip that many coins. If you flip 100, you might end up with only 48, or some other number, of them coming up heads.

F: Well, yes, but I know that if the number of flips gets very large, it is more and more likely that the proportion of them coming up heads will approach 50 %.

B: What do you mean by "more likely"? All you have is some faith in a belief about coin flips, which is exactly my point of view.

"Quantum Bayesianism," developed by Christopher Fuchs and others, follows this to a logical conclusion and says that the probability in quantum mechanics is also a belief; that is, the wave function, which when squared gives the probability of an event, is really representative of a belief; that is, it exists only in the mind of the person, called an agent, who is viewing the universe. The universe does exist, and the agent can operate on it, getting the results of an experiment, and thus changing the wave function, for example, by collapsing it. The collapse of a wave function occurs only because the agent gets new information and she must then update her wave function, that is, what she believes about the system being studied.

Wigner's friend becomes a good example, of this altered view of the wave function. Wigner's friend opens the box containing Schrödinger's cat and notes the cat is alive. Before opening the box, her wave function for the cat was a superposition of alive and dead, as shown in Eq. (8.2). When she opens the box and sees the cat is alive, her wave function for the system collapses to a single term having cat alive and atom undecayed, because she has new information. Now, Wigner comes into the room; when he was outside and did not know the result of opening the box, his wave function was the entangled superposition given in Eq. (8.4). As soon as he sees what the experimenter found, his wave function also collapses to a single term. The point is that there is no single physical wave function that exists external to the agents involved; each has his or her own wave function, if they have different belief sets. Each can change their probability of events by gathering new information, which can occur by an experiment or even by talking to someone who knows the result of the experiment. According to QBists, quantum mechanics is a "user's manual" for operating in the world; it instructs us with what probabilities we should place our experimental bets. But a quantum state is not a representation of the world.

An even more complicated and instructive example is the one involving Alice and Bob doing a Bell experiment with a singlet spin state. When the particles were emitted, they were in an entangled state of up/down minus down/up. When Bob does his experiment and finds an up spin, then his wave function for the other spin predicts down. Alice's wave function does not collapse until she makes her measurement. In the QBist view, there is no problem with nonlocality or instantaneous correlation, because the collapsing of each wave function is a separate local event, each occurring when each agent makes his or her measurement. According to Bob, his wave function collapses to a single term at his measurement. Suppose Bob phones Alice to tell his result before she does her spin-up/spin-down measurement. Even though quantum mechanics says the outcome is certain in her experiment, her wave function is still a matter of her belief. One might ask, as she walks over to her spin apparatus, whether a pre existing outcome is already there. The QBist denies this and says that when Alice makes her measurement, "something new comes into the world that was not there before." The outcome comes into existence by the measurement.

QBism does confirm the existence of an external world, but it rearranges how we think about its relation to quantum mechanics and to the agents interacting with the external world. It is interesting to compare this theory to a statement made by Schrodinger:

> One can only help oneself through something like the following emergency decree: Quantum mechanics forbids statements about what really exists–statements about the object. Its statements deal only with the object-subject relation. Although this holds, after all, for any description of nature, it evidently holds in a much more radical and far reaching sense in quantum mechanics.
>
> Erwin Schrödinger, 1931 letter to Arnold Sommerfeld

## 8.6 Tests of the reality of the wave function

Recent research has been trying to test the *reality* of the wave function, that is, in terms we have used earlier in this chapter, to determine whether an ontic interpretation or the epistemic interpretation is best for quantum mechanics. In the epistemic QBist view discussed in Sec. 8.5, the wave function represents one's *knowledge* of the underlying physical state of a system and thus can vary from person to person, depending on what that person knows. Thus, Wigner's wave function can collapse to that of his friend, if the friend gives him the latest data. The wave function in this view is not directly a measure of reality. In classical probability theory, this sort of thing happens all the time. An airplane goes down in the ocean; crash investigators draw a map of its possible positions on the ocean floor. The map will span, say, a thousand square miles to cover most likely resting places. The plane is in only one actual place, but the probability map extends over a wide area, based on the information the investigators have. In the opposite ontic view, each wave function represents a separate real physical state of nature rather than one person's information of the state.

If one believes that the wave function measures information rather than representing a precise actual physical state, one seems to be assuming there is a further depth to physical reality *beyond* the wave function. When the wave function collapses upon measurement, the experiment is revealing a more detailed picture about that deeper reality, which is mirrored in our increased knowledge of it. When we discussed hidden or additional variables in connection with the Bell theorem, we were talking about a possible deeper hidden reality. The experimental violation of Bell's theory did not say that such hidden variables were impossible, but only that one could not have them in a *local* theory; one measurement can affect another very far away.

The Bohm model discussed in Sec. 5.4.2 interprets the Schrödinger equation in terms of additional variables. The theory describes real particle paths, only one of which is actually traveled by a particle in a performance of a one-particle experiment. The particle is guided by a force that is established by the wave function—a "quantum force." This force limits where the actual particle trajectories might be so that, for example, an interference pattern might emerge after many particles are measured. What puts randomness in the experiment are the initial conditions of the various particles determining which paths are actually traversed. In this case, the wave function gives a measure of the probability of the many possible particle trajectories, but the reality is in each individual actual path. This interpretation of quantum mechanics is perfectly consistent and gives the same results as normal quantum mechanics. But the reality is considered to be the actual particle paths

# Tests of the reality of the wave function 81

(the hidden variables) plus the quantum force and not just solely the wave function.

Suppose we have a wave function $\psi_A$. A Bohmian interpretation will have a certain set of particle paths, $\{\lambda_A\}$, associated with this wave function, any one of which might be followed in one actualization of the experiment. A second wave function, $\psi_B$, might have a completely separate set of particle paths, $\{\lambda_B\}$. Then, we might suspect that the superposition

$$\Psi_{AB} = \frac{1}{\sqrt{2}}(\psi_A + \psi_B) \qquad (8.12)$$

has the joint set $\{\lambda_A, \lambda_B\}$. (This is *not* the way the Bohm model works, as we will explain in the next paragraph.) Assuming this, when we do the experiment and find the wave function collapses to, say, $\psi_A$, we might say that the actual physical state was one of the $\lambda_A$ paths and that it was simply lack of knowledge that made us consider the $\lambda_B$ paths as being possible candidates for the real particle state. Then, $\Psi_{AB}$ was assumed only because of this lack of knowledge and, as just a measure of our state of mind, is not a real physical entity.

However, the Bohm model works differently in its description of the two-slit experiment. When only slit A is open, the Bohm model gives a set of paths, $\{\lambda_A\}$, by which a particle moves through slit A; if only B is open, the set is $\{\lambda_B\}$. However, when one opens slit B when A is already open, all the possible particle paths change instantaneously to completely new paths—$\lambda_{AB}$, which show interference effects. It is a highly non local process. Thus, the particle paths involved with $\psi_{AB}$ are independent of what they were in $\psi_A$, and the two wave functions represent quite separate real situations. Can we decide more generally, beyond just the Bohm model, whether each wave function, even that of a superposition, represents a separate reality?

We assume that there is indeed some hidden-variable reality beyond standard quantum variables. Not every physicist would agree to this assumption, but let's make it. We can continue to refer to the variables as particle paths, although they may be something quite different in different theories and in different experiments. So, to each wave function $\psi_A$, there corresponds a set $\{\lambda_A\}$ of some "paths" whose individual choices in experiments yield the probabilities found in quantum experiments. In any single experiment, only one of these $\lambda_A$ paths is activated, but, over all experiments with that wave function, each of the $\lambda_A$ paths is activated according to some probability $\mu_A(\lambda_A)$. The question is whether a superposition $\Psi_{AB} = \frac{1}{\sqrt{2}}(\psi_A + \psi_B)$ has its $\lambda_{AB}$ paths in common with some or all of the $\lambda_A$ paths, that is, whether the probability distribution $\mu_{AB}$ overlaps $\mu_A$. If there is overlap, then the $\lambda_A$ paths are a reality for *both* $\psi_A$ and $\Psi_{AB}$, and when one of these overlap $\lambda_A$ variables is activated, the question of whether the wave function is $\psi_A$ or $\Psi_{AB}$ is a matter of lack of knowledge rather than a true physical difference. The measurement that collapses the wave function from $\Psi_{AB}$ to $\psi_A$ is one that simply changes our knowledge rather than changing the actual physical state. Proving overlap would require us to accept an epistemic view of nature.

To test these ideas we look at two different MZI devices (see Chap. 6) in Fig. 8.4. Our discussion is similar to an argument given by Lucian Hardy. In the left panel, we have a full MZI called MZAB, in which an input particle is allowed to follow either (or both) arms of the interferometer. Then, the wave function in the arms is just that of Eq. (8.12). Recall that the result of this experiment is that detector D1 triggers but detector D2 does not, because of the interference of the two parts of the wave function at beam splitter 2. In the right panel, which shows the second device, which is called MZA, the lower path B has been inactivated by replacing beam splitter 1 with a mirror, so only arm A is used by the particle. In this case, the beam splitter operation makes the particle equally likely to trigger D1 or D2.

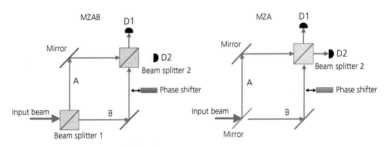

**Figure 8.4** Mach–Zehnder interferometers. *Left*: MZAB: the fully operational interferometer, with two beam splitters. Particles can travel on both arm A and arm B. The interference at the second beam splitter results in particles reaching only detector D1. *Right*: MZA: an interferometer with arm B disabled by replacing beam splitter 1 with a mirror. Here, both detectors D1 and D2 are triggered equally often. A phase shifter can be inserted in arm B and will change the interference behavior at D1 and D2 in MZAB; it will have no effect in MZA.

# Tests of the reality of the wave function

For the interferometer on the right, wave function $\psi_A$ is the correct description, since only arm A is available. If this wave function involves real hidden paths, which are represented by $\lambda_A$, then some of them must "guide" the particle to detector D1, call them $\lambda_A^{D1}$ paths, and others must go to detector D2, call them $\lambda_A^{D2}$ paths. These two kinds of hidden paths occur equally likely. The $\lambda_{AB}$ hidden paths that occur in the full MZI (shown in the left panel) always guide the particle to D1 and never to D2. We can consider the possibility that the $\lambda_A$ hidden paths associated with $\psi_A$ are included in the collection of $\lambda_{AB}$ paths. However, since paths leading to D2 do not occur in the full MZI, no $\lambda_A^{D2}$ paths can be among the $\lambda_{AB}$ paths. But perhaps $\lambda_A^{D1}$ paths are shared, since particles do get to D1 in MZAB.

But now let's change the experiment. We put a phase shifter into arm B of the MZAB. What this phase shifter does is equivalent to increasing the arm length by one-half wavelength, so that, subsequently, a crest appears where a trough would have been, and a trough where a crest would have been. This makes the interference at beam splitter 2 such that particles now only get to detector D2, with none going to D1. We can insert the same phase shifter into arm B on MZA; the wave function $\psi_A$ is unchanged by this, since it describes particles only in arm A. It makes sense to assume the corresponding $\lambda_A$ paths are also unchanged; these include the $\lambda_A^{D2}$ paths. The MZAB wave function $\Psi_{AB}$ is altered after the phase shifter to become $\Psi'_{AB}$. All the $\lambda_{AB}$ paths corresponding to the original wave function $\Psi_{AB}$ must have been changed at the phase shifter to $\lambda'_{AB}$ paths, since they all now lead to D2. So it is possible that the $\lambda_A^{D2}$ paths are now among these. But if they were *all* changed, $\lambda_A^{D2}$ paths could not have been in the original set of $\lambda_{AB}$ paths that existed before the phase shifter was reached or before it was put in. Thus, the set of $\lambda_A$ paths is completely independent of the set of $\lambda_{AB}$ paths, and the two wave functions $\Psi_{AB}$ and $\psi_A$ are completely separate entities with no common underlying physical $\lambda$ variables. Each is representative of its own set of $\lambda$ variables, and so each is real.

There are a lot of assumptions in this argument, and they might seem to make it less than rigorous. Recently, Matthew Pusey, Jonathan Barrett, and Terry Rudolph claimed to prove a theorem (the PBR theorem) that quantum states are real, that is, not just a measure of information, based on proving that any overlap of the underlying physical states (that is, the $\lambda$ variables) would conflict with the predictions of quantum mechanics in measurements on entangled states of a pair of systems. However, one of the assumptions they made, that two physical systems prepared independently have independent physical states, was questioned because it assumed local causality. Since we know that quantum experiments on Bell's theorem infer nonlocality, so that two measurements, even those very far apart, are not necessarily independent, then one cannot assume that two systems, even at great distances, can be prepared independently.

In another approach, theorists have shown that if the quantum overlap between states like $\psi_A$ and $\Psi_{AB}$ is due completely to the overlap of their $\lambda$ distributions, then a certain quantity $S$ satisfies $S \geq 1$. Experiments performed on photons show that this inequality is violated. This implies that the quantum state is real, rather than just being a measure of our uncertain knowledge of the system, although some partial overlap is not yet excluded by experiment.

Many other arguments have been given that seem to favor the ontic models with individual wave functions describing real physical states, rather than epistemic models in which the wave function is a measure of a state of knowledge. However, the discussion is based on assumptions about $\lambda$ distributions, hidden-variable descriptions of a deeper reality. The Bohm model is a very good hidden-variable ontic model. However, not all models of quantum behavior even admit that hidden variables like the $\lambda$ variables are possible. The Copenhagen view questions whether one can make any statements about reality beyond what the wave function predicts or the experiments show. But it lacks any explanation of wave function collapse. The QBist view is epistemic, explains wave function collapse by maintaining that the wave function is just a measure of our knowledge of a real universe, and also denies the existence of any hidden variables describing that reality. The Everett theory (Sec. 8.2) is ontic and has no collapse of the wave function at all. The GRWP theory (Sec. 8.4), in which wave function collapse is a dynamical process governed by an altered Schrödinger equation, is ontic with real wave functions.

# 9

# Bose–Einstein Condensation and Superfluidity

*The crowd makes the ballgame.*

Ty Cobb

In 1924 the Indian physicist Satyendra Nath Bose sent Albert Einstein a paper deriving the thermodynamics of a photon gas by using a new statistical analysis for the thermodynamics of identical particles. Einstein translated it into German and sent it to a prominent German physics journal, which published it on Einstein's recommendation. Einstein realized that the same statistical method could apply to massive particles, not just photons. So, in 1925 he published a paper in which he found that, below a certain temperature, a large number of particles in an ideal gas (one without interactions between the particles) would fall into the lowest quantum energy state of the system; in the case he treated, this was the state of almost zero momentum. As the temperature in an ideal gas is lowered further, more of the atoms condense into this lowest state until, at absolute zero, all of the particles are in the state. Nothing much came of this discovery until much later.

In a seemingly unrelated set of experiments in Leiden in the Netherlands in 1908, Kamerlingh Onnes succeeded in liquifying helium at a temperature of $-452\,°F$. He never mentioned anything special about the behavior of this liquid, although he must have noticed some unusual things. Mostly, he used the liquid to cool other materials in order to study them. In particular, he found that liquid mercury at these temperatures had its electrical resistance vanish; it became a "superconductor."

It wasn't until the 1930s that the unusual properties of liquid helium became evident. Below a certain temperature, it becomes a superfluid: it can flow through very fine channels without friction; it conducts heat extremely well so that, below the transition temperature, boiling stops—no hot spots that cause bubbles can develop; the liquid flows up the sides of a container and might even flow out of it, leaving the container empty; and heating the fluid in a certain way can cause a thin stream of liquid to spray upward, in a "fountain effect." Some of these properties are shown in a series of YouTube videos; the first video in this series is at: http://www.youtube.com/watch?v=OIcFSHAz4E8. In 1936 Fritz London suggested that the effect predicted by Einstein, called the Bose–Einstein condensate, might explain the wondrous properties of liquid helium.

This idea was very controversial for many years. Liquid helium is hardly the ideal gas that Einstein had described; its atoms interact strongly enough that it does liquify, although only at a very low temperature. It wasn't until 1995 that atoms of gases of other elements, mainly alkalis like lithium, sodium, rubidium, etc., held at very low density so that they are almost ideal (non interacting), were found clearly to undergo Bose–Einstein condensation. We will discuss Bose–Einstein condensation in both helium and the alkalis, but first we need to talk a bit about low temperature and about fermions and bosons.

## 9.1 Temperature

Temperature is something for which we all have an intuitive concept. When it is high, we feel hot, and when it is low, we feel cold. It is often, but not always, a measure of the random motion of the molecules or atoms in a substance. That fact would seem to indicate that there could be a lowest temperature where all motion stops. We know from discussions in Chaps. 2 and 3 that quantum systems in their lowest energy state still have zero-point energy, so the motion could not completely stop. But there is indeed a lowest temperature known as "absolute zero." The Kelvin scale is designed to have its zero there (0 K) corresponding to $-459.7\,°F$ or $-273.2\,°C$. Water freezes at $32\,°F = 0\,°C = 273.2$ K. Water boils at $100\,°C = 212\,°F = 373.2$ K. Table 9.1 compares the three scales.

There is a theorem in thermodynamics that says one can never really reach absolute zero; one's refrigerator stops working first. But very low temperatures can be reached. Liquid helium liquifies at 4.2 K and goes superfluid at 2.2 K. However, methods using lasers, magnetic fields, and radio waves have, in recent years, allowed experimenters to cool atoms as low as 0.5 nK (nanoKelvin) = $0.5 \times 10^{-9}$ K, or 1/2 of a billionth of a Kelvin.

# Temperature

**Table 9.1** Three standard temperature scales compared. The Kelvin scale is defined so that absolute zero is 0 K. The Celsius scale is also called the centigrade scale.

|  | Fahrenheit | Celsius | Kelvin |
|---|---|---|---|
| Boiling point of water | 212 °F | 100 °C | 373 K |
| Very hot day | 105 °F | 41 °C | 314 K |
| Freezing point of water | 32 °F | 0 °C | 273 K |
| Very cold day | −40 °F | −40 °C | 233 K |
| Boiling point of liquid air (nitrogen) | −320 °F | −196 °C | 77 K |
| Absolute zero | −460 °F | −273 °C | 0 K |

Liquid helium is cooled by methods somewhat analogous to those used in normal refrigerators, in which certain gases or fluids are forced through a small nozzle. The emerging gas is much colder than the entering fluid because it requires energy, which comes from the thermal energy of the gas, to separate the bound atoms. In the case of helium, a compressed gas enters the nozzle, and an expanded cooled gas exits. Ultimately, the gas cools enough to liquify. To reach lower temperatures, one uses a vacuum pump to remove the hotter gas atoms that evaporate from the surface of the liquid. To get the helium much below 1 K requires other, more sophisticated methods.

Atomic and optical methods of cooling are required to get to the extreme low temperatures required to reach Bose–Einstein condensation for gases. These atomic gases become so cold that they must be kept away from any material walls, which are mostly at relatively very high temperatures, to avoid heating the gas or generating the recombination of atoms into molecules. So, the gases are trapped by magnetic or laser fields. Magnetic trapping requires that the atoms have spins and appropriate energy levels (so-called hyperfine states). Laser methods used for cooling gases also require the use of certain elements that have atomic energy levels that correspond to available laser frequencies. For this reason, it turns out that some of the most appropriate elements are the alkali atoms: lithium, sodium, potassium, rubidium, cesium, etc. Helium and hydrogen have also been used.

Laser cooling methods are rather remarkable since one intuitively expects a laser to heat things up. Since the temperature of a gas is related to the average kinetic energy of the atoms, slowing the atoms cools them. When an atom is moving toward a laser beam, the photons bounce off the atom, slowing it down. But the laser has to be tuned to the frequency at which the atoms scatter most strongly but without absorbing the photon. If the atom is moving toward the laser, the atom sees a slightly higher frequency, since it sees more wave peaks per second (the Doppler effect). The laser is tuned to scatter strongly at this frequency. When the atom is moving away, it sees a lower frequency that is further out of resonance, so that the atom does not scatter the photon and thus is not sped up. As the atom slows down and the gas cools, that frequency must be raised so the photons keep the cooling process going. One has to have lasers incoming from all directions to produce cooling effective in all three dimensions.

Other lasers or magnets provide a force field (usually described by a harmonic potential) that traps the atoms. The potential energy corresponding to the trapping force can be adjusted to allow atoms with an energy greater than some set value to escape. Thus, very hot atoms (i.e., those having high kinetic energy) will escape. The remaining trapped atoms will, on average, have a lower temperature, having been cooled

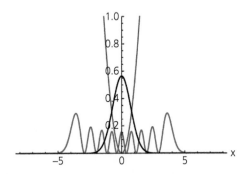

**Figure 9.1** Two harmonic oscillator states representing atoms trapped in a magnetic or optical potential. The harmonic potential is shown in gray; the ground state probability function ($n = 0$) is in black, and that for $n = 8$ is in red. The average energy of the ground state is 0.5 units, while that of the $n = 8$ state is 8.5 units. If the harmonic potential were cut off, as shown, at 1.0 and not allowed to go any higher in energy, the $n = 8$ state would escape the trap. Letting higher energy states escape cools the remaining gas.

## 9.2 Fermions and bosons

The basic building blocks of the atom are protons, neutrons, and electrons. Protons and neutrons bind together in a small, heavy clump known as the nucleus, while the lighter electrons, attracted electrically to the protons, are spread in a much larger cloud around the nucleus. Schematic diagrams of the two isotopes of helium are shown in Fig. 9.2. $^4$He has a nucleus with two protons and two neutrons and which has a radius of about 2 femtometers ($2 \times 10^{-15}$ m). Two electrons are farther out from the nucleus but should *not* be thought of as being in orbits like planets around the sun. They are much more delocalized in wave function clouds. The electronic radius of helium is 30 picometers ($30 \times 10^{-12}$ m), more than ten thousand times larger than the nucleus. Hydrogen is the simplest element with one proton and one electron. A two-dimensional representation of the electronic wave functions is given in Fig. 9.3.

Isotopes are atoms with the same number of protons in the nucleus and the same number of electrons (attracted electrically to the protons), but differing numbers of neutrons in the nucleus. The chemical behavior of an element is mostly determined by the electrons. Helium at high temperatures is an inert gas; it is not active chemically. There is a rarer helium isotope, $^3$He, which has only one neutron in the nucleus

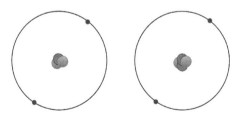

**Figure 9.2** A crude schematic illustration of the two isotopes of helium. $^3$He has a nucleus with two protons (blue) and one neutron (green), while $^4$He has a nucleus with two protons and two neutrons. The two electrons (red) in each atom are *not* actually in orbits as shown here but are smeared as described by non localized wave functions. See the more realistic depiction of a one-electron wave function in Fig. 9.3.

90    Bose–Einstein Condensation and Superfluidity

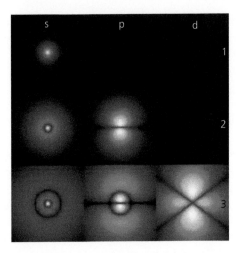

**Figure 9.3** A representation of hydrogen atom electronic states. The states are three dimensional, but only their two-dimensional projections are shown. Each is identified by three quantum numbers $(n, l, m)$, where $n$ is the number on the right, and $l$ is represented by the letters s, p, and d. Higher states with larger $n$ values and more $l$ values are not shown; $m = 0$ here. The energy depends only on the so-called radial quantum number $n$. The other quantum numbers ($l$ and $m$) represent the angular momentum states of the electron. (Figure credit: Florian Marquardt, Erlangen.)

(the superscript on He is the atomic mass number, which indicates the total number of nucleons in the nucleus). $^3$He and $^4$He have very different thermodynamic behaviors at low temperature! We need to see why.

All particles come in two types: *fermions* and *bosons*. Elementary particles such as electrons, neutrinos, muons, and quarks are fermions. They all have spin 1/2. Composite particles made up of an odd number of fermions are also fermions. Thus, protons and neutrons, which are each made up of three quarks, are fermions with spin 1/2. Any atom with an odd number of protons, neutrons, and electrons, for example, $^3$He, is a fermion.

Fermions obey what is called *Fermi–Dirac statistics*. Two identical fermions, say, two electrons, one with wave function $\psi_a$, and the other with wave function $\psi_b$, must have a joint wave function that looks like this:

$$\Psi(1,2) = \frac{1}{\sqrt{2}} \left[ \psi_a(1)\psi_b(2) - \psi_b(1)\psi_a(2) \right]. \tag{9.1}$$

Because one electron looks and behaves exactly like another—they are identical particles—we cannot tell if they interchange places. So, we must entangle the states, as shown. The state is required to be antisymmetric, that is, if we interchange the electrons, the first and second terms become interchanged, and the wave function becomes minus what it was. We formed a similar wave function when in developing the singlet spin state in Eq. (5.8). Only there, we considered an electron and a positron, which are non identical particles; in that case, we could also have a triplet state, in which interchange would leave the same wave function with no change in sign. Suppose that, in Eq. (7.1), $a = b$, that is, the states of the two electrons are identical; then, the wave function vanishes! This feature is called *the Pauli principle*: only one electron is allowed in any single-particle state.

Electronic energy states similar to those shown for hydrogen (see Fig. 9.3) exist for nuclei with more protons. Helium has two protons, which electrically attract the two electrons. There are $s$ states for the electrons in the helium atom, somewhat like those for hydrogen. Now, only one electron is allowed in each state, but the electron has a spin, so that the two electrons both go into the $s$ state but with opposite spins: one up, and one down. In atoms with three electrons, the third electron would go into the higher energy 2s state. Boron has two electrons in the 1s state, two in the 2s state, and one in the 2p state. Each of the higher energy states has a larger radius, and heavier atoms are correspondingly larger. The helium atom is chemically inert because it has "closed shell" of 1s states. Without the Pauli principle, all electrons could fall into the lowest state, the s-state, and matter would collapse. Chemical reactions would be impossible, and life would not exist. Moreover, one atom avoids another because when the electron clouds overlap, the Pauli principle provides a mechanism for a strong electrical repulsion between the atoms. Matter occupies space; I can pick up a rock, and my hand does not pass right through it! We will have more to say about fermions in Chap. 11.

Particles with integral spin, for example, 0, 1, 2, etc., are bosons and obey *Bose–Einstein statistics*. Among these are photons, gluons (particles that cause the force that holds the quarks together in the proton and neutron), the Z and W particles that mediate what is called the weak interaction, and the famous Higgs boson, the recently discovered particle that provides mass to all the elementary particles. Also, composite particles containing even numbers of fermions obey Bose–Einstein statistics. These particles have symmetric wave functions. Two particles in the same state $\psi_a$ would have the wave function

$$\Psi(1,2) = \psi_a(1)\psi_a(2). \quad (9.2)$$

If the two particles are in different states, so now we have $\psi_a$ and $\psi_b$, then

$$\Psi(1,2) = \frac{1}{\sqrt{2}} \left[ \psi_a(1)\psi_b(2) + \psi_b(1)\psi_a(2) \right]. \quad (9.3)$$

An interchange of the particles leads to the identical wave function with no sign change; that is, the wave function is symmetric. There is a Bose principle analogous to the Pauli principle; bosons *prefer* to be in the same state. This is the principle that leads to Bose–Einstein condensation.

## 9.3 The laser

The laser is a good illustration of the Bose principle, which states that bosons prefer being in the same state. In a laser, atoms are each "pumped" into an excited state, a quantum state with energy above the lowest one. If one atom drops to the lowest state, it emits a photon; this photon "stimulates" a nearby atom to emit a photon in exactly the same state: not only at the same frequency (because that is determined by the electron energy state) but also with the same phase and direction (see Fig. 9.4). Then, the two photons stimulate a third, and a fourth, and

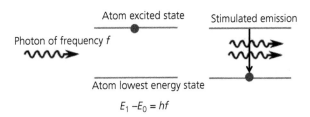

**Figure 9.4** Stimulated emission of a photon, caused by the presence of an initial photon in a laser. The two photons stimulate other atoms to emit photons all in the same direction and with the same phase. The photons are reflected by mirrors at the ends of the laser cavity so they can stimulate more atoms to emit photons. Some escape through a semi reflecting mirror in one of the ends to form a laser beam; $E_0$, energy of the atoms's lowest energy state; $E_1$, energy of the atom's excited state; $f$, frequency; $h$, Planck's constant.

there is an avalanche of photons into the same state. Mirrors at the ends of the tube reflect the photons back and forth so that, with continual pumping of the atoms into excited states, the laser beam builds. A semi reflecting mirror allows some of the light to escape, giving a very bright, coherent beam. This is a case of a Bose effect in which multiple photons end up in the same state.

## 9.4 Superfluid helium

The most abundant isotope of helium, $^4$He, with two protons, two neutrons, and two electrons, has an even number of fermions and so is a boson. Two helium atoms have a force of attraction between them, due to the combined electrical effects of the electrons and protons. (If the atoms get too close to one another, the force becomes a repulsion, as discussed in Sec. 11.2.3.) This attractive force is fairly weak, so the gas does not liquify until a very low temperature, 4.2 K. One might expect that as one lowers the temperature still further, the liquid would finally solidify like any other material, as all thermal motion is removed. But it does not; the force is so weak that the quantum zero-point energy is enough to keep it liquid even at absolute zero. However, one can confine the atoms manually in place by applying pressure to squeeze the system into a smaller volume, so that, at about 25 times atmospheric pressure, the atoms do form a solid. The resulting solid behaves weirdly enough (e.g., quantum tunneling is evident) to be called a "quantum crystal."

At 2.2 K the liquid becomes a superfluid; it can flow without friction. In 1936 Fritz London suggested that the cause of this transition was Bose–Einstein condensation, in which a substantial portion of the atoms fell into their lowest energy state, which was the (almost) zero-momentum state. His idea has since been confirmed. Bose–Einstein condensation begins at 2.2 K in liquid helium, with one particle in the zero-momentum state stimulating other atoms to enter this state. As the temperature is lowered, more atoms fall into this lowest state. At absolute zero, about 8 % of the atoms will be in the lowest state; interactions between the particles prevent further occupation. Nevertheless, the system becomes superfluid. To understand how a liquid can flow without friction, we need to understand what causes friction between a flowing fluid and, say, the walls of a tube through which it is flowing. Fig. 9.5 illustrates a fluid flowing past a small object immersed in

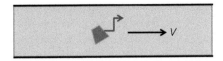

**Figure 9.5** An object flowing through a normal fluid loses energy and momentum, which constitutes friction by causing an excitation of atoms (the crooked red arrow) in the fluid—thus transferring its energy to the fluid. There are fluid states of very low energy that are easily excited so friction happens at any speed $v$ of the object. A fluid flowing past a fixed object will have excitations in the direction opposite to its flow, causing a loss of momentum.

the fluid. Instead of considering the fluid flowing past the object, we consider the equivalent case of the object moving through the liquid. The object loses energy and momentum by creating excitations of the fluid, possibly giving energy to individual atoms or groups of atoms that remove energy from the object. A fluid flowing around a fixed object will receive similar excitations that oppose its motion, slowing it down.

But, in superfluid helium, the bosons prefer to move together as a large unit. Single particle states, in which an atom might move on its own, are not present. When one atom moves, a whole group of atoms wants to go along. In essence, this makes the excitations that cause friction have large mass and high energy. Thus, there is an energy gap in the system, as shown in Fig. 9.6. If the object in the fluid is flowing slowly enough, it is unable to create an excitation with enough energy. Thus, it

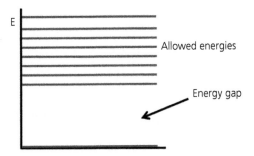

**Figure 9.6** The Bose principle causes atoms to move in a coordinated manner so they move together rather than as individuals. The result is an energy gap, so an object moving in the liquid cannot transfer small amounts of energy to the fluid. Thus, if the object is moving slowly enough, it flows without friction.

flows without friction. However, there is a critical velocity above which it can excite a quantum state of the liquid beyond the energy gap, so the superflow ceases.[1]

If we put liquid helium in a container with a bottom that is a very fine filter, with extremely small channels, normal helium fluid (above 2.2 K), cannot flow through it because the fluid is viscous (has internal friction). But once the temperature falls below that temperature, the fluid goes through as if the filter had wide channels—it loses its viscosity. Until absolute zero, not all of the fluid is superfluid; there is still some normal fluid present. As the superfluid flows through the filter, the normal fluid is left behind but is cooled quickly to produce more superfluid. This then flows through the filter, and the process continues until all the fluid in the container has passed through the filter.

The "creeping film" effect in helium is explained in the same way. Every fluid will coat the sides of any container it is in; the walls attract atoms to it. However, in a container with an open top, the superfluid coating the surface can began to flow, and friction will not stop it. The film on the surface acts like a siphon, and the fluid flows up the walls and out of the container until the container is empty.

A film can be easily formed on a glass ring, and the fluid can be induced to flow around the ring in a circular path. The velocity of the film can be monitored, and it is found that in principle, it would last permanently; the film does not lose energy. This experiment was done by Robert Hallock in 1974 at the University of Massachusetts Amherst.

## 9.5 Bose–Einstein condensation in dilute gases

Because the atoms in liquid helium are interacting rather strongly, only 8 % of the atoms fall into the lowest state, even at absolute zero. The more dilute the system is, the less frequent are the atom–atom interactions. Moreover, the strength of the interaction can be altered in some dilute systems. So if one could make a dilute gas undergo Bose condensation, it would be possible to have almost 100 % of the atoms fall into the lowest state. This was the motivation for using the alkali gases: these gases are naturally liquids or solids at normal temperatures but

---

[1] To be more technically accurate, there some quantum states in the energy gap; these are sound wave states, but for particular reasons, they cannot be excited in the flowing fluid or in an object moving in the fluid.

can be maintained in the gas phase on a fairly long timescale while their temperature is lowered, as long as they stay dilute. So one can gather quite small numbers of particles, say several thousands to several millions, in a magnetic or optical trap that forms a harmonic potential, much as we studied in Chap. 3. The transition temperature depends on the density. As the temperature is lowered, the average atom energy is lowered, and the quantum wavelength increased. When the wavelength becomes equal to the average distance between particles, the boson nature of the atoms becomes evident, and Bose condensation occurs. With the very low density we are considering, the transition temperature might be only 200 nK (nanoKelvin), but the methods described in Sec. 9.1 allow these low temperatures to be reached.

Fig. 9.7 shows two graphics of gases as their temperatures are lowered from above the transition temperature to below. The condensate is only a few tens of microns (millionths of a meter) wide and so is difficult to image. To make these figures, the experimenters turn off the potential to let the gas expand and then take "shadow" images, that is, shine laser light through the gas. The 3D plot is made to peak up where the density is largest. Note that there is a tight condensate in the middle, with atoms in the wider excited states surrounding the central condensate. The condensate, in the top figure, is not symmetric because the original trap was not. The density has the basic shape of a harmonic oscillator ground state probability function.

The characteristic of a Bose–Einstein condensate is that many atoms are in the lowest energy harmonic state Bose–Einstein condensate and those atoms act in a very coherent way. One of the most amazing experiments is the one performed by Wolfgang Ketterle's group at MIT. They divided a condensate into two parts and then let the condensates move into each other. Because the condensates are matter waves, they interfere and form the pattern shown in Fig. 9.8. This result shows the wave nature of matter in a most spectacular way, with the matter waves interfering alternatively constructively and destructively, causing oscillations in the density of particles.

These condensates can be used to show quantum mechanical effects in many ways. They are also useful for making accurate atomic clocks, possible "atom lasers" (coherent beams of particles), systems for accurately measuring rotation, etc. They can show matter interference in other ways in addition to that shown in Fig. 9.8. For example, one can make a particle interferometer that is similar to the Mach–Zehnder

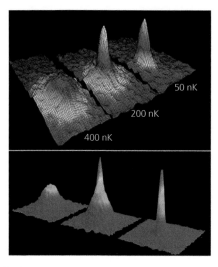

**Figure 9.7** *Top*, shadowgrams of rubidium gas at 400 nK (above the transition temperature), 200nK, and 50 nK. The high-energy states are wider than the ground state wave function and surround the condensate density. This condensate was imaged by the JILA group (the joint institute of the University of Colorado at Boulder and the National Institute of Standards and Technology) under the direction of Carl Wieman and Eric Cornell. The peaks in the graph are made to be proportional to the density and are artificially color coded as well. *Bottom*, similar plots from the MIT lab of Wolfgang Ketterle. Here, the number of atoms was 700,000, with a transition temperature of 2 μK (microKelvin). Wieman, Cornell, and Ketterle shared the 2001 Nobel Prize for their work. (*Top*: Used with permission. Credit: Michael Matthews, JILA. *Bottom*: Used with permission. Credit: MIT Ketterle group)

photon interferometer shown in Fig. 6.1. A Mach–Zehnder matter-wave interferometer was constructed by a German group and dropped from a 120 m tower. The interference pattern seen is shown in Fig. 9.9. The reasons for dropping the device are several, including, for example, ultimately to study the effects of Einstein's theory of gravity on quantum objects. Here, the free fall resulted in the matter waves traveling long distances to reach the final interference, thus making the device effectively hundreds of times bigger than a stationary interferometer. Thus, it was converted into a macroscopic quantum object.

In Fig. 9.10, we present a complicated interferometer device. Without going into the details, we simply note that the device is designed to

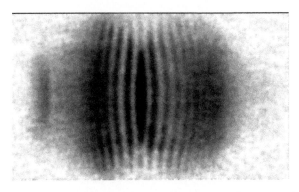

**Figure 9.8** Interference between two overlapping condensates. Where the wave functions interfere destructively, the density is low (light shadow); where there is constructive interference, the density is high (dark shadow). From M. R. Andrews, C. G. Townsend, H.-J. Miesner, D. S. Durfee, D. M. Kurn, W. Ketterle, Science 275, 637–641 (1997). Reprinted with permission from AAAS.

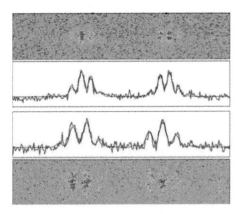

**Figure 9.9** A Bose–Einstein condensate microgravity experiment in which a matter-wave Mach–Zehnder interferometer was dropped from a 120 m tower found the interference counts in each detector, as indicated in the figure. (Reprinted with permission from H. Müntinga et al, Phys. Rev. Lett. 110, 093602 (2013). Copyright 2013 by the American Physical Society.) Thanks to Jakob Reichel, Ècole Normale Supérieure, Paris, for the figure.

produce a superposition of two states in arms 5 and 6: in the one state, all $N$ of the particles are in arm 5, with none in arm 6; in the other state, all the particles are in arm 6, with none in arm 5. For large $N$, this is equivalent to a Schrödinger cat state—a macroscopic object in a superposition of two very disparate states. One might call arm 5 the "alive"

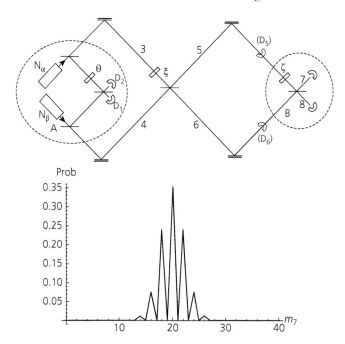

**Figure 9.10** *Top*, a complicated theoretical interference device to create a Schrödinger's cat state, which in this case is a superposition of two states, one of which has almost all $N$ bosons in arm 5, and the other of which has almost all $N$ in arm 6. *Bottom*, a predicted interference pattern between the two cats when $N = 80$. The graph gives the probability of finding $m_7$ particles in detector 7. (From W.J. Mullin and F. Laloë, Jour. Low Temp. Phys. 162, 250 (2011). Copyright 2011. With permission of Springer.)

state, and arm 6, the "dead" state. By using this device, we could see the predicted interference between the two parts of the superposition. So far, however, this device is purely theoretical.

There are many phenomena that occur in Bose–Einstein condensates but which we have not considered, for example, quantized vortices, which are basically microscopic whirlpools; these also exist in superfluid helium. One of the most productive uses of Bose–Einstein condensates is in simulating more complicated systems. In some cases, theoretical predictions can be made for a condensed matter system, but the experiment with the original system is too complicated to perform or it is hard to untangle the phenomenon sought from other effects there. So, one produces a simpler condensate analog for the original real system. This allows testing theoretical models where the various effects can be more easily separated from one another. For example, Stephen Hawking had predicted in 1974 that black holes could emit particles allowing it to lose energy (Hawking radiation will be discussed in Chap. 13.) A recent theoretical analysis has shown the possibility of seeing the mathematical equivalent of Hawking radiation, in this case, sound waves (phonons), in a flowing Bose condensed gas.

There are other superfluids we have not considered. In metals and ceramics, electrons (fermions) can pair to form a kind of boson known as a Cooper pair. These condense to form a superconductor, in which electrons can flow without electrical resistance. Electromagnets formed by coils of superconducting wires form the basis of all MRI machines in hospitals. The materials become superconducting only at low temperatures, so each MRI machine contains liquid helium, which is used as the coolant. We mentioned in Sec. 9.2 that liquid $^3$He behaves quite differently from liquid $^4$He. Indeed, liquid $^3$He can have fermion pairing to form a superfluid, but this occurs at a temperature a thousand times lower than that in the boson $^4$He isotope, namely, at a few milliKelvins. The pairing of very dilute fermion gas particles in the optical traps we described above also leads to analogous fermion superfluids. Low temperature physics is full of interesting quantum objects.

# 10

# The Quantum Zeno Effect

> Every line is the perfect length if you don't measure it.
>
> MARTY RUBIN

The ancient Greek philosopher Zeno of Elea (490–430 BC) presented a paradox that has Achilles, a very fast runner, racing to catch up with a tortoise. Achilles gives the tortoise a head start. By the time Achilles has covered that ground, the tortoise has moved on some distance. By the time Achilles has covered this new distance, the tortoise has again moved, and so on. Whenever Achilles gets to where the tortoise has been, the tortoise has already moved on, so it seems that Achilles can never catch up.

Another of Zeno's paradoxes involves an arrow that has been shot from a bow. At any instant of time, an arrow is at a single place; but, says Zeno, if it is at only a single place, it must be at rest, so motion is impossible. Zeno seemed to be supporting the philosophic position that existence is timeless and unchanging and that motion is an illusion of our senses.

Both these paradoxes involve the assumption that time is made up of instants, each of which has no duration. The effects we discuss are also related to the adage "a watched pot never boils." In quantum mechanics, if we watch a system constantly to see it when it moves to a new state, it just won't move to that state. It becomes like Zeno's arrow.

## 10.1 Measuring an atom's energy level

To see how this happens, consider an atom in an electromagnetic field, such that photons of the right frequency cause the atom to jump from its ground state, state 1, to a higher state, state 2. When the atom is in the higher state, stimulated emission will take place, causing the atom to return to state 1. (We have discussed stimulated emission in relation

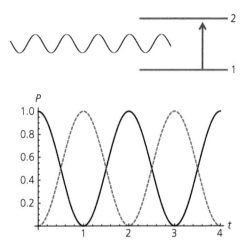

**Figure 10.1** *Top*: An atom with two energy levels is irradiated with photons having energy matching the difference between levels 1 and 2. The photon is absorbed, and the atom starting out in level 1 goes to level 2. In level 2, it is stimulated by the radiation field to emit a photon and drop back to level 1. *Bottom*: The probability of the atom being in level 1, as a function of time, is shown by the solid black line; the probability of it being in level 2 is shown by the dashed red line. The curves oscillate with the so-called Rabi frequency $\Omega$, here taken as $\pi$ per second.

to the laser in Sec. 9.3.) See Fig. 10.1(*Top*). It turns out that the probability of finding the atom in energy level 1, which is where it starts at time $t = 0$, is given by

$$P_1(t) = \cos(\Omega t/2)^2, \tag{10.1}$$

while the probability $P_2$ of finding it in level 2 is

$$P_2(t) = \sin(\Omega t/2)^2. \tag{10.2}$$

The quantity $\Omega$ is called the Rabi frequency and depends on the strength of the interaction between the photon and the atom. These probabilities are plotted in Fig. 10.1(*Bottom*). They add up to unity, as they should.

We start the atom out in the lowest state, state 1, and, after some time interval, determine the state of atom. The probability $P_1$ that the atom remains in the lowest state is almost 1 if we observe the atom very soon after the start, and $P_1$ drops off until it is 0 at time $t = T = \pi/\Omega$.

# Measuring an atom's energy level

Meanwhile, the probability of the atom being in level 2 has grown to unity. Suppose we measure the atom's energy state at time $T/2$. The probability of finding it in level 1 is $\cos(\Omega T/4)^2 = \cos(\pi/4)^2 = 1/2$. Suppose we find it has remained in level 1; we now measure it again after the same time interval. The probability $\mathcal{P}(2)$ that it is still in the lowest state after these two measurements is the probability that it was there after the first measurement, times the same thing again, or

$$\mathcal{P}(2) = \left[\cos(\pi/4)^2\right]^2 = \frac{1}{4} = 0.25. \tag{10.3}$$

Suppose now we make four such measurements, with each being at the smaller interval $T/4$. Then, the probability of the atom still being in the ground state is

$$\mathcal{P}(4) = \left[\cos(\Omega T/8)^2\right]^4 = \left[\cos(\pi/8)\right]^8 = 0.53. \tag{10.4}$$

We have looked more often at the state and the probability of finding it still in the lowest state is now larger. If we look at shorter time intervals $T/n$ where $n$ is larger than 4 we find

$$\mathcal{P}(n) = \cos(\pi/2n)^{2n}. \tag{10.5}$$

This probability is plotted in Fig. 10.2, and we see that as the number of measurements increases, with the time between them getting shorter, the probability of the atom staying in the ground state goes to 1, that is,

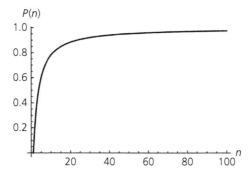

**Figure 10.2** The probability of an atom being in its lowest state after time $T = \pi/\Omega$ when measured every time interval $T/n$.

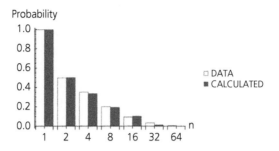

**Figure 10.3** The probability of an atom making the transition from level 1 to level 2 after being measured $n$ times in the interval $T$, as measured by the Wineland group. Note that this is $1 - \mathcal{P}(n)$, so it is large at small $n$, and zero at large $n$. (After Itano et al, Phys. Rev. A 41, 2295 (1990)).

it never goes to the excited state: this is the quantum Zeno effect. If we had not measured the atom, it would certainly have been in the excited state after the total time $T$. Similarly, if it started in the excited state, it would stay there if continually measured at infinitesimally short time intervals.

We have not said how we would actually determine whether the atom is in the ground state. However, in 1990, the experimental group of David Wineland, who won the 2012 Nobel Prize for his work with cold ions, figured out how to do this. It involves looking at a third energy level that is strongly connected to level 1. One shines laser light at the atom to excite it from level 1 to level 3. If these photons are absorbed and reemitted, then the atom was in level 1; if you see no photons, then it was in level 2. The result of these experiments and the calculation of the probability are shown in Fig. 10.3.

## 10.2 Rotating polarization

Recall our description of polarization of photons in Chap. 6. One can start with a polarization in a particular direction. But there are liquids and solids that can rotate the angle of polarization—they are "optically active." For example, if the photon enters a sugar solution with its polarization pointing up and down, it might exit with it at a 10° angle with the vertical. (The polarization direction is rotated because the sugar molecule has "chirality," a handedness, like a screw. Sugar companies use this property to measure the concentration of sugar in a solution.) Suppose we line up several such polarization-rotating

# Rotating polarization

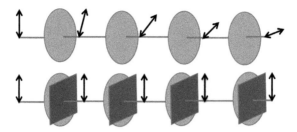

**Figure 10.4** *Top*: Vertically polarized photons pass through an optically active material that rotates the polarization by $22.5° = \pi/8$ each time. After four passes, the polarization is now horizontal. *Bottom*: Here, after each polarization rotation of $\pi/8$, the photon hits a vertical polarizer, so it has a probability of $\cos^2(\pi/8) = 0.854$ of getting through each. After four such passes, the probability of exiting with vertical polarization is $(0.854)^4 = 0.53$.

elements and watch photons go through, as in Fig. 10.4(*Top*). As the photon goes through, its polarization is rotated a bit ($90°/4 = 22.5° = \pi/8$ in radians in the figure) each time until it exits the series with horizontal polarization. But suppose we interpose a vertical polarizer at each stage, as in the bottom panel of Fig. 10.4. We saw in Sec. 6.2 that this would allow only vertical polarization through. For a photon at a polarization angle $\alpha$ with respect to the vertical, the polarizer would, according to the Malus law, allow the photon through with probability $\cos(\alpha)^2$. If $\alpha = 0$, the photon definitely gets through; if $\alpha = \pi/4$ (45°) it gets through 1/2 the time.

Suppose that there was just one optically active device, and it rotated the angle by 90° ($\pi/2$ in radians). If we put in the polarizer, the probability of the photon getting through is $\cos(\pi/2)^2 = 0$. That makes sense, since the polarization of the photon is perpendicular to the polarizer orientation. Next, we interpose two rotators, each of which rotates by $\pi/4$, and two vertical polarizers. The probability that the photon gets through is

$$\mathcal{P}(2) = \cos(\pi/4)^4 = 0.25, \tag{10.6}$$

which is small but no longer zero.

If we put in $n$ of these combinations of rotator and polarizer, with each rotation being by an angle of $\pi/2n$, the probability of getting through is

$$\mathcal{P}(n) = \cos(\pi/2n)^{2n}. \tag{10.7}$$

The corresponding probability for the situation shown in the figure, with $n = 4$, is 0.53. This is the same function we plotted in Fig. 10.2, and it goes to 1 at large $n$. That is, if we keep rotating the polarization and then letting it through with probability corresponding to that ever-decreasing rotation angle, we don't stop the photon at all; it goes through unimpeded. The group of Anton Zeilinger did an experiment equivalent to this but using mirrors to allow a photon to pass through the same rotator and polarizer many times.

## 10.3 Bomb detection: The EV effect

Related to the quantum Zeno effect is what is called "interaction-free measurement" or "counterfactual computing." This is a case where quantum superposition makes possible something so weird that one might think it is equivalent to extrasensory perception. Avshalom Elitzur and Lev Vaidman invented this technique, so it is called the EV effect. To make this look fancy, they concocted a story about someone, call him Mr. X, who had a collection of bombs, each of which had a very sensitive detonator, such that if a single photon hit it, the bomb would explode. However, some of the bombs were duds, and the detonator would not set those off. Now, Mr. X needed his bombs to be reliable for his intended use (whether good or evil). So, how is he to tell? If he tries one, it will not explode if it is dud, and it can be put aside. But if it is good it explodes when he tries it, and he no longer has that one! This looks like an impossible situation, but EV come to his rescue with their quantum bomb detector. This is quantum weirdness at its weirdest.

We set up an MZI (see Fig. 10.5). When working normally, with no bomb, the first beam splitter causes the photon to enter a superposition of states in the two arms of the interferometer. These interfere at the second beam splitter such that they have constructive interference at D1, and destructive interference at D2. So, one always gets a count only in D1. But now suppose the bomb is put in with the mirror acting as the detonator. A dud will cause a reflection of a photon, without explosion, and the result is a normal count at D1, and none at D2. One should repeat this test a few times to make sure that all counts are in D1 and never in D2. Now, suppose the bomb is working; then the bomb will provide an observation of which arm the photon takes. If it goes in the lower arm, the bomb explodes, and we lose that bomb. But if it goes in the upper arm, we can get a count in either D1

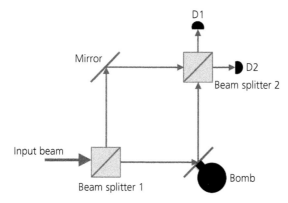

**Figure 10.5** The Mach–Zehnder interferometer arranged for bomb detection. When acting normally with just a mirror and no attached bomb, a photon enters at the lower left and beam splitter 1 puts it into a superposition of wave functions in both arms, which meet again at beam splitter 2. Here the two beams interfere to produce a count in detector D1, and *none* in detector D2.

or D2. If we get it in D2, we know the bomb is a good one, and we put it aside for later use. However, there is an equal chance that beam splitter 2 will deflect a photon in the upper arm into D1, and we will not know if we have a good bomb or not. So, whenever we get a D1 count, we must repeat the experiment. If the bomb is good, we will ultimately get either an explosion or a count in D2, both of which tell us the bomb is good. Thus, we can detect a good bomb by using a photon *that does not touch it*. This is the explanation for the term "interaction-free measurement."

With this particular bomb detecting method, we lose a lot of bombs. It turns out 33 % of the good bombs will be successfully found with the device, with the rest exploded. But the group led by Anton Zeilinger showed how one could increase the efficiency greatly. Consider a series of very many interferometers, as shown in Fig. 10.6(a). The beam splitters are set to allow only a small portion of a beam to get through, or, in other words, a single photon has a small probability of getting through the beam splitter and is most often reflected. But calculations (that are a bit too complicated to give here) show that the photon gradually transfers from the upper set of paths to the lower set. After an infinite number of attempts, it would have a 100 % probability of being in the lower set. Interference effects are very important in this transference.

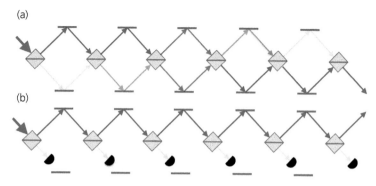

**Figure 10.6** (a) A photon in a multiple interferometer with beam splitters that have a high probability of reflecting the photon. The probability of the photon being on any path is indicated by the shading of the path; here, the photon has a low probability of being in the lower arms at the beginning but gradually becomes more and more likely to be there. (b) Multiple interferometers with absorbing detectors on the lower paths. The beam splitters highly favor reflection, so the photon mostly stays in the upper path, with some small likelihood of being detected in the lower arms. The more reflections there are, the more likely it is that the photon will exit via the upper path.

Now, consider the lower set of interferometers in Fig. 10.6(b), with each beam splitter again having a high probability of reflecting the photon. If it goes through any one of the beam splitters, it is detected and stops. So if it is able to continue, it is always in the upper set of paths. We then arrange to have the probability of reflection be $\cos(\pi/2n)^2$ (that same cosine function again!) where $n$ is the number of beam splitters. When $n$ is large this is very close to 1. The probability of it being in the upper paths after $n$ reflections is

$$\mathcal{P}(n) = \cos(\pi/2n)^{2n}, \tag{10.8}$$

a familiar relation plotted in Fig. 10.2. We see that, after a very large number of beam splitters, this probability approaches unity; the photon is very unlikely to end up in a detector and so will exit via the upper path.

Suppose we can turn the detectors on or off, with "off" meaning that the photon passes on in the lower paths and hits the lower mirrors, as in panel (a). If they are off we find that the test photon exits on the lower path. But if they are on, the photon definitely exits on the upper path. But then, we have detected the on position without ever having touched a detector—an interaction-free measurement. You know they are there without "seeing" them!

Next, we remove all the black detectors and instead connect every lower level mirror by some electrical wiring to a single bomb. If any one of the lower path mirrors is hit, a good bomb will explode—a good bomb is equivalent to a detector. If a photon leaves by the lower path, we know the bomb is a dud (detectors off), but if it leaves by the upper path, we know the bomb is good (detectors on). We have detected the good bombs with almost 100 % efficiency by using an interaction-free measurement. Bizarre!

A real interaction-free measurement was done by the Zeilinger group to show that the effect works. Fig. 10.7 shows a diagram of the apparatus. Suppose the moveable mirror is to the left, out of the photon path. After the beam splitter the photon is in a superposition of two paths: one to the mirror on the bottom, and the other to the mirror on the right. This results in interference such that detector D1 does not click and the photon goes back out the way it came in. With the moveable mirror in position, there are three possibilities: (1) D1 clicks, (2) D2 clicks, or (3) the photon again goes back out the entry path. But if D1 clicks, we know the moveable mirror is in position without the photon

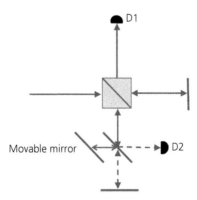

**Figure 10.7** An interaction-free measurement device. A photon enters from the left and can either be reflected downward or continue to the right through the beam splitter. With the moveable mirror out of position, the two paths interfere so that there is destructive interference for detector D1, and the photon passes back out the entry port. With the moveable mirror in position, the photon can click detector D1 or D2 or exit by the entry path.

having gone that route. If it had gone that route, D2 would have clicked. The theory fits the experimental data very accurately.

There have been attempts to use interaction-free measurement in an imaging device in which photons do not touch the object imaged! This could have valuable uses. For example, some biological molecules are very sensitive to radiation; if x-rays were used to image them, the molecules would be broken up. In Fig. 10.6, we saw that when an object was in the lower path, as in panel (b), the photon took the upper path; when there was no object to absorb the photon in the lower path, it took that path, as in panel (a). This suggests the possibility that we could vary the lower path systematically so that sometimes it intercepts part of the object, and other times it does not. We can tell where the parts of the object are by the path taken. By varying the paths, we map out the object with no photon ever touching the object.

Very recently, researchers at the Austrian Academy of Science and the University of Vienna, produced an actual image (the silhouette of a cat) by a slightly different method, in which photons passing through the object were never detected, and the image was formed by photons that never touched the object. The key to doing this was that the detection probabilities of the observed photons depended, because of interference, on the transmission probabilities and phase shifts of the photons passing through the object. Thus, the detected photons could be used to form an image because they carried the same information as those actually "seeing" the object. This could allow, for example, the use of probe wavelengths for which efficient detectors are not available, since the probe photons are not detected! The cover figure to this book shows an image produced by this method.

# 11

# Bosons and Fermions

> Why did the chicken cross the road? There already was a chicken on this side of the road.
>
> WOLFGANG PAULI

Back in Sec. 9.2, we mentioned that all elementary particles come in two general categories: bosons and fermions. We also showed that this resulted in very distinct wave functions for the two types—symmetric for bosons and antisymmetric for fermions—and that this difference resulted in very different behaviors for the two types of particles. Indeed, we showed that bosons can have Bose–Einstein condensates and superfluidity. Here, we investigate this distinction in more detail and see other evidence for how the basic fermion or boson character is manifested.

## 11.1 Wave functions

Let's review some things we discussed previously. We noted that elementary particles such as electrons, neutrinos, muons, and quarks are fermions; more complicated tightly bound particles, like nuclei and atoms, that contain an odd number of fermion elementary particles are also fermions. The proton, for example, which is made up of an odd number of quarks, is a fermion. The helium isotope $^3$He, which has two protons, one neutron, and two electrons, is also a fermion. In contrast, photons, gluons, Z particles, and W particles are bosons, as is the Higgs boson. Particles made up of even numbers of fermions, such as mesons (made up of a quark and an antiquark) and atoms like the $^4$He atom, which has two protons, two neutrons, and two electrons, are also bosons.

Bosons obey *Bose–Einstein statistics*, which means the wave function does not change when we interchange the two particles in the wave function. This feature arises because the particles are indistinguishable and we cannot tell which one is in which state. Two bosons in the same state $\psi_a$ would have the wave function

$$\Psi_{\text{Bose}}(1,2) = \psi_a(1)\psi_a(2), \tag{11.1}$$

which is properly symmetric: there is no change when we interchange particles 1 and 2. If the two individual states $a$ and $b$ are different, then

$$\Psi_{\text{Bose}}(1,2) = \frac{1}{\sqrt{2}}\left[\psi_a(1)\psi_b(2) + \psi_b(1)\psi_a(2)\right]. \tag{11.2}$$

Interchanging particles 1 and 2 again leads to no sign change, but now we must have two terms. In a sense, the bosons *prefer* to be in the same state. This is the principle that leads to the Bose–Einstein condensation we discussed in Chap. 9.

Fermions obey *Fermi–Dirac statistics*. Two identical fermions, say, two electrons, one with wave function $\psi_a$, and the other with wave function $\psi_b$, have a joint wave function that looks like this:

$$\Psi_{\text{Fermi}}(1,2) = \frac{1}{\sqrt{2}}\left[\psi_a(1)\psi_b(2) - \psi_b(1)\psi_a(2)\right]. \tag{11.3}$$

The state is required to be antisymmetric; that is, if we interchange the electrons, the first and second terms become interchanged, and the wave function becomes minus what it was. Because probability is the square of the wave function, this change of sign makes no difference in the probability. But the identical nature of the particles means one cannot say for sure which one is in which state, so the wave function must have both terms when $a$ and $b$ are different. On the other hand, if the two states $a$ and $b$ are the same, the two terms in the wave function cancel, and the wave function vanishes, yielding the Pauli principle that keeps ordinary matter from collapsing. This principle keeps two fermions from being in the same state and often seems to indicate a kind of repulsion between two Fermi particles, but this view is too simplistic, as we will see in Sec. 11.2.

## 11.2 Examples of Fermi and Bose effects

### 11.2.1 Average particle separation

Let's look at a very simple calculation of the distance between two particles. Suppose the particles are in states $a$ and $b$ but are not identical;

they are distinguishable–perhaps one is a $^3$He atom, and the other a $^4$He atom; when we interchange the two particles, we can see the change has been made, unlike the case where the two are identical. We assume for now that the there are no interparticle forces. The wave function for distinguishable particles is simply

$$\Psi_{\text{Dist}}(1,2) = \psi_a(1)\phi_b(2), \tag{11.4}$$

where $\psi$ describes the $^3$He atom, and $\phi$ describes the $^4$He atom. There is nothing to keep $a$ and $b$ from being the same state in this case, say, the same momentum. The square of the average separation of the two particles in this case is

$$\langle(x_1 - x_2)^2\rangle_{\text{Dist}} = \langle x_1^2\rangle_a + \langle x_1^2\rangle_b - 2\langle x_1\rangle_a \langle x_1\rangle_b, \tag{11.5}$$

where $\langle x_1^2\rangle_a$, for example, means the average of the square of the first particle's position in state $a$. This involves squaring the $\psi_a(1)$ wave function and adding up and averaging over all the possible positions that it allows. (Mathematically, this involves what is called an integral in calculus.) When we do the same for fermions and bosons by using Eqs. (11.2) or (11.3), we find that the average distance between the particles is the distinguishable value, plus a quantity $X$ for fermions, and minus the same value for bosons. For the record,

$$\langle(x_1 - x_2)^2\rangle_{\text{Fermi}} = \langle(x_1 - x_2)^2\rangle_{\text{Dist}} + X, \text{ and} \tag{11.6}$$

$$\langle(x_1 - x_2)^2\rangle_{\text{Bose}} = \langle(x_1 - x_2)^2\rangle_{\text{Dist}} - X, \text{ where}$$

$$X = 2(\langle x_1\rangle_{ab})^2,$$

and where the average in $X$ is over the positions allowed by the product $\psi_a(1)\psi_b(1)$; here, in a sense, the particle is in both states in this term. We have run into this kind of "exchange term" before involving the overlap of two different wave functions of a superposition and causing interference. So, two fermions spend more time apart and bosons spend more time together than distinguishable particles do. This makes it seem like there is an extra force of attraction between bosons and a repulsion between fermions, although there is no actual force in our model.

### 11.2.2 The second virial coefficient

Suppose we have helium gases at high temperature. The helium atom comes in two forms, as we have mentioned: $^3$He, a fermion, and $^4$He, a boson. At normal temperatures, these substances are gases; their liquefaction occurs at low temperature: 4 K for $^4$He, and 3 K for $^3$He. Not surprisingly, the pressure of the Fermi gas is slightly lower than that of the Bose gas when they are at the same temperature and have the same number of particles. The reason is that multiple fermions cannot all be in the same momentum state, and so they are spread out to higher momenta, causing harder hits, on average, to the walls, and higher pressure. In contrast, multiple bosons can exist in the same momentum states and so can have lower overall average momenta and hit the walls less hard, on average. The simplest correction to the famous ideal gas law $PV = NkT$ (where $P$ is pressure, $V$ is volume, $T$ is temperature, $N$ is the number of atoms in the gas, and $k$ is the "Boltzmann constant") is called the "second virial coefficient" and adds a positive term to the right side for $^3$He, and a negative term for $^4$He. Quantum effects can be seen at quite high temperatures via this correction. Such "statistical" quantum corrections have to be sorted out from second virial corrections due to real interactions between the particles, as these lower the pressure, if the particles attract one another, or raise the pressure, if the particles repel each other. Quantum effects also occur in these interaction terms.

### 11.2.3 Interatomic forces

When two helium atoms (either isotope) are sufficiently far apart, such as they would be in the liquid state on average or in a gas, they attract one another. This so-called van der Waals interaction is electrical in nature. While the electron cloud around the nucleus is usually symmetrical, there are fluctuations in which the nucleus and the electron cloud's center become separated from one another. Such a "dipole" (slightly separated positive and negative charge) induces a dipole in another nearby helium atom, and the two charge distributions attract one another. However, when the atoms are crowded too close together, then the electrons are being pushed into the same positional state, and the Pauli principle forbids this. The electron density then becomes reduced in the region between the nuclei, which can now feel their mutual positive charge repulsion more strongly. This repulsive atomic force at small atomic separation is what keeps matter from collapsing.

### 11.2.4 White dwarf stars

A similar thing happens when a star of sufficiently small mass burns out its nuclear fuel, and the hydrogen atoms have all fused into helium atoms and heavier atoms. The star collapses into what is called a white dwarf star. Gravity is what makes the star collapse when the energy released by the nuclear reactions no longer holds it at its usual star diameter. But the same effect that caused the second virial effect, that is, the necessity of electrons being in higher momentum states because of the Pauli principle, results in a "Pauli pressure" that opposes the shrinking effect of gravity. If the star is massive enough (about 1.5 times the mass of the sun), then the Pauli pressure is inadequate to oppose gravity, the collapse continues, and a "neutron star" is formed. If the mass is even greater (about 10 solar masses), it continues on to become a black hole.

## 11.3 Inclusion of spin

When we include spin in the makeup of the particles, our wave function discussion is basically valid but needs to be refined a bit. Fermions are particles with spin 1/2, as we discussed in Chap. 9; each one has what appears to be a spinning charge allowing it to behave as if it were a small magnet. In addition, as we know, each spin can be in one of two quantum states; up or down. Consider the following possible wave Fermi function for two particles:

$$\Psi_{\text{Fermi}} = \frac{1}{\sqrt{2}} \left[ \psi_a(x_1)\psi_b(x_2) + \psi_b(x_1)\psi_a(x_2) \right] \quad (11.7)$$
$$\times \frac{1}{\sqrt{2}} \left[ u_+(1)u_-(2) - u_-(1)u_+(2) \right].$$

The first part of the equation describes the spatial state of the two particles, which can be in states $a$ and $b$ at positions $x_1$ and $x_2$. This might describe the momentum states of the particles and leads to the probability of finding the particles at certain positions. The second part describes the spin states of the two particles: in the first term, particle 1 has spin up (+), and particle 2 has spin down (−), with the opposite in the second term. If we interchange particles 1 and 2, we have to exchange both the space and the spin variables simultaneously. When we do that, the space factor in the wave function does not change—the second term becomes the first, and the first, the second. However, the spin term changes sign, so the overall wave function is properly antisymmetric. Now suppose the $a$ and the $b$ states are the same; then, the space factor should be simply $\psi_a(x_1)\psi_a(x_2)$ (with the spin factor still taking care of the antisymmetry), which is like a boson spatial wave function! The two particles are not in exactly the same state, because the spins are different, with one fermion up, and one down. The lowest electronic energy state of the two electrons in the helium atom is approximately of this form.

On the other hand, we can also construct a two-particle fermion wave function that has the two spin states the same; the spatial part must be antisymmetric then. One possibility is this:

$$\Psi_{\text{Fermi}} = \frac{1}{\sqrt{2}}\left[\psi_a(x_1)\psi_b(x_2) - \psi_b(x_1)\psi_a(x_2)\right] \quad (11.8)$$
$$\times \frac{1}{\sqrt{2}}\left[u_+(1)u_+(2)\right].$$

Here, both spins are up. The wave function for both spins down simply has the plus signs replaced with minus signs. Can you write out a fermion wave function in which $a$ and $b$ are different but which has an antisymmetric spatial wave function, with one spin up and the other down?[1]

### 11.3.1 Polarization methods

Several years ago, researchers discovered excellent ways to make highly polarized systems, that is, systems with most of the particles having their spins pointing in the same direction. In one case in Paris, at the École Normale Supérieure, physicists invented a way to do "optical pumping" of atoms, that is, putting them in excited states. One main result was the laser, in which such excited atoms collapse simultaneously to the lowest energy, producing coherent light. (See Sec. 9.3.) Alfred Kastler won the Nobel Prize in 1966 for inventing optical pumping. The technique has a side benefit; it can be used to polarize $^3$He

---

[1] Interchange the plus and minus signs in Eq. (11.7) to get

$$\frac{1}{\sqrt{2}}\left[\psi_a(x_1)\psi_b(x_2) - \psi_b(x_1)\psi_a(x_2)\right] \times \frac{1}{\sqrt{2}}\left[u_+(1)u_-(2) + u_-(1)u_+(2)\right].$$

**Figure 11.1** Three generations of Nobel Prize winners: photo taken at the École Normale Supérieure in 1966, when Alfred Kastler won the Nobel Prize for the development of optical pumping. Also appearing are Claude Cohen-Tannouji, winner of the prize in 1996, for theoretical contributions to laser cooling, and Serge Haroche, winner in 2012, for "measuring and manipulation of individual quantum systems." Jean Brossel was a close associate of Kastler, and the laboratory is named after both. (Photo copy from Franck Laloë, ENS, Paris.)

atoms in a gas, as shown in 1963 by Forrest Colegrove, Jr., Laird Schearer, and Geoffrey Walters. Given this tool, scientists began looking for interesting things to do with the polarized gas. We will get to what they suggested below. See Fig. 11.1, which shows a photo taken at the École Normale Supérieure in 1966, showing three generations of Nobel Prize winners!

Spins are magnets, and so an external magnetic field will cause them to line up. A method of producing polarized $^3$He liquid involves polarizing the solid with a strong magnetic field; this is fairly easy because, in a solid, the atoms are all in different spatial positions, so the spatial wave function of any pair can be antisymmetric (and still not vanish, since $a$ and $b$, representing different positions in the crystal lattice, are not the same), and the spin states can all be the same, as in Eq. (11.8). But when the solid is melted, with most of the spins pointing up, say, all the new spatial states, which involve differing momenta, will have very high energies to keep the particles from having the same spatial state. One has a high polarization, but it will not last very long because the spatial energy can be made smaller by have spins flip over so that as many are down as are up, and pairs of momenta can then be the same. The polarization is "metastable," that is, will last only for some "relaxation time." However, this time is long enough that some experiments on polarized liquid $^3$He can be done.

A third way to polarize $^3$He is to make it very dilute by mixing it into liquid $^4$He where it floats around like a thin gas. A strong magnetic field can then polarize it. Because of the diluteness, the antisymmetry of the spatial part does not cost much extra kinetic energy.

Incidentally, the ability to reach high polarizations has some interesting medical applications. An MRI machine works by polarizing spins (usually protons) in one's body and looking at specific properties of the spins, such as the time to relax to equilibrium. The spin properties depend on the local environment, and an image of these variations is produced. The higher the polarization, the better is the image. Presently, very large magnetic fields made using superconducting magnets (cooled by liquid helium) are used to produce the polarization.

### 11.3.2 Transport processes

**Thermal conductivity**

Suppose I heat the end of a copper bar to 100 °C and hold the other end in ice water at 0 °C. Heat will flow from the hot end to the cold end. By measuring how fast the ice melts, I can determine the rate at which heat flows. For a given temperature difference, the amount of heat flowing along the rod also depends on the length, with the shorter bar conducting more heat per degree temperature difference. It also depends on the cross-sectional area of the bar, with a wide bar conducting more heat than a narrow one. One can measure the "thermal conductivity" in watts per meter degree Celsius. (The watt is a unit of energy per second). In a metal like copper, mobile electrons carry the heat energy. (They can carry electric current as well.) Copper is found to be a much better conductor of heat than, for example, steel, wood, or most other substances. Superfluid helium, in which quantized sound waves carry the heat, has a thermal conductivity even better than copper. Now, suppose we measure how well polarized $^3$He atoms in a gas can conduct heat. The rate of energy flow is impeded by collisions between the atoms. When one atom has a lot of energy and is traveling from hot to

# Inclusion of spin 119

cold, it soon smacks into another colder atom and transfers much of its energy to the second atom. The shorter the path to the next collision, the lower is the thermal conductivity. If the spins of a pair of atoms are both up (as symmetric spin wave function), then the space wave function has to be antisymmetric, as in Eq. (11.8) and must become small when two atoms are close; note that it vanishes at $x_1 = x_2$. The two atoms don't even get close enough to interact with one another; they simply do not collide! Only if the spins are different can they interact, but, in a polarized system the number of these pairs is small. The result is that polarizing the system causes the thermal conductivity to become extremely large.

The common idea that there might be an extra force of repulsion between two Fermi particles because of the Pauli principle is contradicted by the conductivity measurement; in this case, the principle actually says that the two polarized fermions don't even see one another.

### Viscosity

Another possible transport measurement is that of viscosity. Suppose I strum a guitar wire in a vacuum; it will vibrate at some frequency—although no sound will be produced. Now, if I vibrate it in air, the air causes a resistance to its motion; some of the air is dragged along with, or pushed by, the motion of the wire. By this interaction of air and wire, the guitar sound is produced. Obviously, the wire in vacuum will vibrate for a lot longer than it will in air. If we put the wire in molasses, its vibration will die out even more quickly. Molasses has a higher "viscosity" than air; a vacuum has zero viscosity. Oil, on the other hand, has a lower viscosity than, say, molasses and can be used as a lubricant. We can measure the viscosity by the changes in the frequency and vibration lifetime of the wire. A wire in a viscous medium has a lower frequency because it is dragging more mass along, and its lifetime of vibration is shorter.

We can measure the viscosity of $^3$He gas or of $^3$He dissolved in liquid $^4$He by using a vibrating wire. The vibrating wire in pure superfluid liquid $^4$He shows a very low viscosity, which is what we mean by calling that substance a superfluid. But if we dissolve $^3$He atoms in the liquid, the viscosity goes up because the wire interacts with them, and the $^3$He atoms interact with each other, causing the increase. If the $^3$He atoms are polarized, they no longer collide; it is the interatomic collisions that cause the viscosity, which now decreases. The wire sees less drag; its frequency increases and the vibration lifetime increases by many factors of ten.

### 11.3.3 Ferromagnetism

Ferromagnetism, the phenomenon in which solids like iron, nickel, and cobalt become permanent magnets, is very complicated. The electrons in a metal are mobile and can be considered to be moving through the entire volume somewhat like a gas or liquid. Here, we present one model of a simple ferromagnetic liquid, although this model does not apply to iron magnets. We suppose we have a liquid with a number of spins with no external magnetic field. If the kinetic energy of a particle is $\epsilon$, and there are $N_+$ up spins and $N_-$ down spins, the kinetic energy of all the spins is $(N_+ + N_-)\epsilon$. But the spins interact with each other, and we must include that energy as well. We assume, as in the transport case, that the spatial wave function of a pair is antisymmetric and that two up spins interact only very weakly; the same holds for two down spins. However, two unlike spins can get close enough to interact and have an interaction energy in the form $gN_+N_-$, that is, depending only on the product of the number or up and down spins. So the total energy is

$$E = (N_+ + N_-)\epsilon + gN_+N_-. \qquad (11.9)$$

If $g$ is positive, corresponding to a repulsion between two like spins, then the lowest energy state will be that with all spins either up ($N_- = 0$) or down ($N_+ = 0$), so that the interaction energy will vanish. The spins will align one way or the other, with nature basically flipping a coin to determine which it is. If the interaction occurred with negative $g$, the energy would favor a situation in which the $gN_+N_-$ is made as negative as possible, that is, with equal numbers of up and down spins, so that $N_+ = N_- = N/2$, producing a kind of "anti-ferromagnet."

## 11.4 The Hong–Ou–Mandel effect

In 1987 C. K. Hong, Z. Y. Ou, and L. Mandel (collectively referred to as HOM) performed an experiment showing an effect that has since been named after them. We illustrate the idea by using the simple interferometer shown in Fig. 11.2. Consider a scenario in which one boson ($N_\alpha = N_\beta = 1$) emanates from each source, with both heading

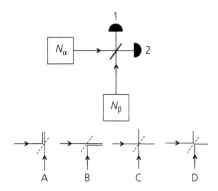

**Figure 11.2** *Top*, an interferometer with one beam splitter and two detectors, 1 and 2. The sources α and β send $N_\alpha$ and $N_\beta$ particles, respectively, to the beam splitter. *Bottom* When $N_\alpha = N_\beta = 1$, there are four possible particle paths, specified by A–D. In A and B, both particles go to one side, with one reflected, and the other transmitted. In C and D, there is one particle found in each detector, with both the particles being reflected or transmitted.

simultaneously to the beam splitter. In this case, there are four possible outcomes, as shown in Fig. 11.2(*Bottom*). The last two, C and D, cancel out by interference in the wave function. The final wave function is thus a superposition of A and B. A measurement always finds two bosons emerging on the same side—bosons like to go together. The experiment used photons and counted the rate of cases where one photon exited on each side. The number of such cases decreased substantially when the two photons exited at identical times, showing the improbability of this event.

This kind of process can be generalized to cases where $N_\alpha$ and $N_\beta$ are not unity. For example, if $N_\alpha = N_\beta = 4$, then the theory predicts that one will always find an even number of particles, that is, 2, 4, . . . , 6, or 8, in each detector, again showing the bosons preference for pairing. The HOM experiment has been repeated successfully for atomic bosons. Note that, in this case, one has to be able to detect and count individual atoms, which is a technology that has been perfected only recently.

## 11.5 The Hanbury Brown–Twiss experiment

There is another famous experiment, done in 1956, that seems to show the essential boson character of photons. Robert Hanbury Brown

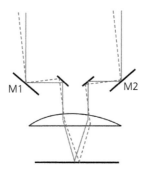

**Figure 11.3** A Michelson interferometer. Light coming in vertically (solid blue lines) interferes constructively at a point on the screen. Light entering at an angle (red dashed lines) has slightly different path lengths for the two beams and can destructively interfere if the path lengths differ by half a wavelength. The bright and dark fringes will be separated enough to be visible only if the mirrors M1 and M2 are sufficiently far apart. For starlight, the angles are greatly exaggerated.

(a double last name) and Richard Twiss (the two are collectively referred to as HBT) were attempting to measure the angular width of stars; this is a very hard task, but angular width is an important property of stars. One method for doing this uses a Michelson interferometer, which depends on the interference ability of light waves. Consider Fig. 11.3, which shows light coming in two possible directions from a single star. The light coming from one direction is reflected off the two mirrors M1 and M2 to other mirrors that guide it to a point through a lens. (A lens is not really used in the actual telescope, but having it simplifies the diagram.) Suppose the telescope is aligned so that the two paths, represented by solid lines, have equal path lengths. They then interfere constructively where they meet if the light sources for the two beams are coherent and in phase. Adjusting the two main mirrors to be aimed at the same spot on the star is vital then. Light from a slightly different angle (the dotted lines) will be slightly off-center, so the paths taken by the two beams will have slightly different lengths and so might interfere destructively. Being able to see both a bright and a dark fringe depends on the star having a sufficient angular width; one can measure the angular width of narrower stars if the distance between the two main mirrors M1 and M2 is made larger. But the problem with large separations of the mirrors is maintaining the coherence of the light all the way to the detector.

# The Hanbury Brown–Twiss experiment

However, HBT had a different idea for this kind of measurement. Light coming from stars is thermal, and it fluctuates, that is, has changes in brightness, because different parts are at different temperatures and intensities; even the same point fluctuates with time. If two light beams are correlated in their fluctuations, they should be from the same region of the star; if they are from far-removed parts of the star, they will be uncorrelated. So, perhaps looking at correlations of light from different regions will be related to the angular width of the star. What HBT saw was so surprising to most physicists that they did not believe it. The explanations led to the development of the science called quantum optics; in 2005 Roy Glauber won the Nobel Prize for his work on the quantum theory of optical coherence, stimulated in part by the HBT experiment.

Consider Fig. 11.4. We are looking for correlations in the arrival of photons from two sources, $a$ and $b$, at two detectors, A and B.

We can calculate the wave function for such simultaneous arrival; it should be of the form

$$\Psi(A, B) = \frac{1}{\sqrt{2}} \left[ \psi_a(A)\psi_b(B) + \psi_b(A)\psi_a(B) \right]. \tag{11.10}$$

That is, we can have the photon from atom $a$ (in a star, say) arrive at detector A ($\psi_a(A)$) while the photon from atom $b$ arrives at detector B ($\psi_b(B)$). But the wave function has to be symmetric, because the photons are bosons. That means we must also consider the other "exchange" term, which has the photon from atom $a$ arrive at detector B ($\psi_a(B)$) while the photon from atom $b$ arrives at detector A ($\psi_b(A)$). The probability will depend on the square of the wave function and have terms like

$$P_{AB} = \Delta_{Aa,\,Bb} + \Delta_{Ab,\,Ba} + E_X, \tag{11.11}$$

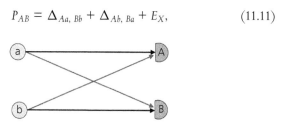

**Figure 11.4** Light from two sources, $a$ and $b$, can reach the detectors in two ways: via the direct paths ($a$ to A, and $b$ to B) or via the exchange paths ($b$ to A, and $a$ to B). Both these must be included in the probability for simultaneous detection in detectors A and B.

where the Δs are "direct" terms, that is, $\Delta_{Aa,Bb}$ is proportional to the square of the first term in Eq. (11.10), and $\Delta_{Ab,Ba}$ proportional to the square of the second term, and $E_X$, the "exchange" term, depends on the product of the two terms: $\psi_a(A)\psi_b(A)\psi_a(B)\psi_b(B)$. This product depends on the time of arrival of the two photons. If that from $b$ arrives at A much later than that from $a$, the product $\psi_a(A)\psi_b(A)$ will vanish. Suppose the two sources are close together and are coherent—corresponding to atoms that emit radiation simultaneously—and the two detectors are also close together. Then, the paths between, say, $a$ and A, and $b$ and A, will be almost the same, and the direct and exchange terms will each correspond to simultaneous hits on the detector A—and on B as well—so the exchange terms contribute as much as the direct terms. However, if $a$ and $b$ are far apart, the path lengths will be too different to have simultaneous detection events; in this case, the exchange term is reduced. Moreover, if the detectors are moved apart, the coincidence between arrivals will diminish because the path lengths differ. The exchange term becomes small then too.

What is plotted in Fig. 11.5 is a correlation function that is adjusted to be 1 when the occurrences are random and not coincidental; only the direct term contributes then. When there is coincidence between

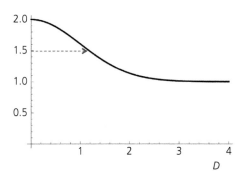

**Figure 11.5** A Hanbury Brown–Twiss correlation function as a function of distance $D$ between the two mirrors M1 and M2. When the distance is large enough, the correlation function drops from 2, which indicates perfect coincidence of the two photons, to 1, which indicates random arrival of the two photons. The width of the function indicated by the dashed arrow depends on the separation of the sources, and the angular size of the star.

the photons reaching either of the two detectors, the direct and the exchange terms both contribute, so the correlation function is twice as big. What we are seeing is that the photons are bunched; photons from different sources tend to arrive together, perhaps as we might expect from bosons. That two photons from different regions of a very hot star could each care where the other one is rather surprising, but that is a characteristic of coherence in quantum mechanics. Unfortunately, in the case of light, the old classical theory, which says nothing about the existence of photons, is able to give a similar result for the correlation function! Thus, the HBT experiment really does not does confirm quantum theory very well; the classical and quantum theories are just consistent.

However, the HBT experiment has been done with atoms by a French group of experimentalists. It turns out one can cool helium atoms that have their electrons in excited energy states to very low temperatures, even to being Bose–Einstein condensed, and so make the atoms suitable for such an experiment. The atoms are cooled in a small magnetic trap, which is then turned off so that the atoms fall to a detector that can record the positions and times of their arrivals. If the atoms are $^4$He, we have bosons, and the experiment is much like the photon experiment. The correlation function should look like the one shown in Fig. 11.5. On the other hand, if we have fermions, the wave function is antisymmetric, with a minus sign instead of a plus sign:

$$\Psi(A, B) = \frac{1}{\sqrt{2}} \left[ \psi_a(A)\psi_b(B) - \psi_b(A)\psi_a(B) \right]. \tag{11.12}$$

In this case, there is antibunching of the particles, and the exchange term, having a negative sign, *cancels* against the direct term, making the correlation function smaller at the origin. Fermi particles avoid coming in simultaneously. This kind of effect is impossible in any classical theory, so this experiment confirms the quantum effects beautifully as we see in Fig. 11.6, which shows the actual experimental data.

The $^4$He experiment just described used temperatures that were higher than the Bose–Einstein condensation temperature. But suppose all the particles were condensed into the same lowest quantum state. Then, states $a$ and $b$ in the wave function would be the same:

$$\Psi(A, B) = \psi_a(A)\psi_a(B). \tag{11.13}$$

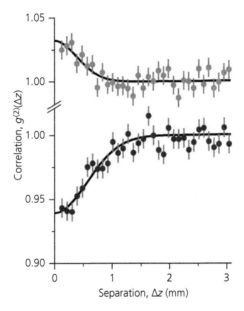

**Figure 11.6** The correlation functions showing the analog of the Hanbury Brown–Twiss experiment done with excited-state helium atoms. The distance $\Delta z$ is a function of the difference in arrival times at the detector. The upper curve shows the effect with $^4$He particles demonstrating the bunching of pairs of particles. The lower curve shows the fermion anti-bunching occurring with $^3$He atoms. In the $^4$He case, the temperature was above the Bose–Einstein condensation temperature. For various experimental reasons, the Bose correlation function does not reach a maximum of 2, and the Fermi one does not vanish at small distances. (Reprinted by permission from Macmillan Publishers Ltd: T. Jeltes et al, Nature 445, 402405. Copyright 2007).

There would be no different $b$ state. Whatever the source, the corresponding wave function would be the same state $a$; thus, there would be no exchange term. The correlation function would be a uniform unity everywhere. And that is what is found in the data!

## 11.6 What more?

There are many other examples where experiments show the basic Fermi or Bose character of particles. It is really particle identity that is involved. Because quantum mechanics—especially the uncertainty

# What more? 127

principle—rather blurs our detailed vision, we cannot tell one particle from another if we exchange them. When we interchange the two particles, the wave function stays basically the same—for fermions, it does change sign, but when it is squared to give a probability, the minus sign cancels out, so the fermion probability stays the same under interchange. However, the two possibilities, symmetric or antisymmetric wave functions, Eqs. (11.2) and (11.3), that result from this effect maintain a huge difference, in the sign of the exchange terms in the two probabilities. This is especially well demonstrated in the remarkable differences in the behavior of the various phases of the two isotopes of helium, $^3$He and $^4$He. The huge difference in behavior is caused by the missing neutron in the nucleus of the $^3$He atom, which results in the substances having the different statistics, as well as a smaller atomic mass and a spin 1/2 for $^3$He. Such remarkable differences pervade physics because of this distinction between bosons and fermions, and it is rather amazing that quantum mechanics is able to account for the huge differences by a simple sign change in a wave function.

# 12

# The Quantum Eraser

> Who controls the past controls the future. Who controls the present controls the past.
>
> GEORGE ORWELL

We have discussed the two-slit experiment as a fundamental example of superposition (see Sec. 4.2). The particle, say an electron, goes through both slits in order to form the interference pattern. If one tries to detect through which slit the electron passes, the interference pattern is ruined. Einstein suggested a possible way to determine which way the electron went as shown in Fig. 12.1. If the electron passes through the top slit, it should interact with the upper barrier attached to a spring and because the electron is deflected a bit, the barrier should rebound, compressing the spring slightly, and similarly if the electron went through the bottom slit. Thus, one could tell which way the electron went. Bohr then pointed out that the recoil would slightly displace the electron's path by an amount $\Delta x \approx \hbar/\Delta p_x$, where $\Delta p_x$ was the recoil momentum of the upper slit in the vertical direction. This amount of deviation was just enough to wash out the interference pattern on the screen. One can either have the wave like interference pattern or the particle like path knowledge; the two are seemingly incompatible with one another in this case.

In Sec. 8.4 on decoherence, we discussed an alternative way of detecting the slit taken by using a camera, as shown in Fig. 12.2. If the camera resolution is adequate, as shown, the images on the photograph will allow one to tell which slit the electron used. We saw that probability for an electron hitting the screen in a particular place in this case is

$$P_{\text{el/ph}} = \frac{1}{2}\left(\psi_1 p_1 + \psi_2 p_2\right)^2 = \frac{1}{2}\left(\psi_1^2 p_1^2 + \psi_2^2 p_2^2 + 2\psi_1\psi_2 p_1 p_2\right), \quad (12.1)$$

where $\psi_i$ describes the electron coming from slit $i$, and $p_i$ represent photon functions at the camera. Since $p_1$ and $p_2$ do not overlap, their

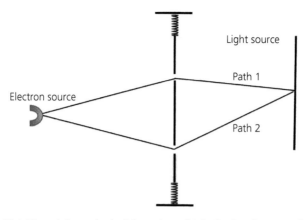

**Figure 12.1** Einstein's method of detecting which slit the electron takes. The recoil of the slit barrier disrupts the interference pattern.

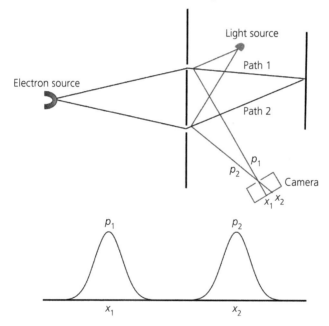

**Figure 12.2** *Top*: A light source and camera are used to determine which slit the electron used. *Bottom*: The camera resolution must be such that the images associated with the two paths are non overlapping, as shown. The interference pattern is destroyed.

product vanishes, and the interference term in $\psi_1\psi_2$ vanishes, even if $\psi_1$ and $\psi_2$ were to continue to overlap. Moreover, one could likely show that the photon interaction with the electron would itself disrupt the interference pattern by altering the electron's path so that the interference between $\psi_1$ and $\psi_2$ would be washed out.

A few years ago, however, Marian Scully, Berthold-Georg Englert, and Herbert Walther (collectively referred to as SEW) showed how one could detect "which-way" information without substantially changing the particle's path, so that one would expect that the interference pattern might be maintained. (This claim of no substantial change in the particle's path has been disputed.) The method allows one to choose, even after the particle completes its trip to the screen, whether one has the which-way information or interference. If you choose to know the's particle's path (or even to have just the opportunity to look at it), the interference pattern is ruined; but if you choose to "erase" the which-way information, the interference pattern is restored! We have, therefore a "quantum eraser."

## 12.1 Erasing with atoms, photons, and cavities

Warning: the algebra in this section is a bit complicated, but it should be worth the effort needed to follow it. If it becomes too much, skip to Sec. 12.2. SEW's idea is that one can construct a radiation cavity and cool it sufficiently that there are no photons in it initially. (A "cavity" has highly reflecting walls so that radiation can exist in it without decay for a very long time. Just as a string held tightly to walls allows only certain standing waves, radiation in a cavity is restricted to a set of resonant frequencies.) An atom can be put into an excited state before passing it through a cavity, so that the atom is almost certain to emit a photon while in the cavity and then exit the cavity in its ground state. SEW show that the atom's path is unaffected by this emission of the photon. Figure 12.3 shows the set up. If one detects the photon as being in one or the other cavity, one knows which path was taken, and this is expected to ruin the interference. Thus, one cannot argue that it is the uncertainty principle that upsets the interference pattern but that just the "knowledge" of the which-way information is enough to destroy the pattern. Let's see how this might happen.

If the atom enters the cavity in its ground state, then nothing happens in the cavity while the atom passes through to the slits; the corresponding probability at the screen remains as usual:

Erasing with atoms, photons, and cavities   131

Atom wave function

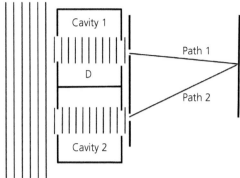

**Figure 12.3** As an atom in an excited state passes through a cavity, it emits a photon; if the photon is in cavity 1, we know the atom went through the top slit; if the photon is in cavity 2, the atom went through the bottom slit. (The vertical lines represent crests of the traveling atom wave function.) D represents a detector that, if used, erases our knowledge of which cavity had the photon.

$$P_{\text{at/ph}} = \frac{1}{2}\left(\psi_1^2 + \psi_2^2 + 2\psi_1\psi_2\right). \tag{12.2}$$

We get our usual interference pattern from the $2\psi_1\psi_2$ term. Suppose, however, that the atom enters the system in an excited state and emerges in its ground state. To treat this case mathematically, we need to include the photon state: if it went through cavity 1, the atom must have emitted a photon there, so the photon/cavity state is $\phi(1_1, 0_2)$ (one photon in cavity 1, and zero photons in cavity 2), and if it took path 2, the photon/cavity state is $\phi(0_1, 1_2)$ (no photon in cavity 1, and one in cavity 2). In Fig. 12.4, we show the wave functions that might represent these photon states. The corresponding system wave function must include the cavity photon as well as that of the atom

$$\Psi_{\text{atom-photon}} = \frac{1}{\sqrt{2}}\left[\psi_1\phi(1_1, 0_2) + \psi_2\phi(0_1, 1_2)\right], \tag{12.3}$$

and the probability becomes

$$P_{\text{at/ph}} = \frac{1}{2}\left[\psi_1^2\phi(1_1, 0_2)^2 + \psi_2^2\phi(0_1, 1_2)^2 \right.$$
$$\left. + 2\psi_1\psi_2\phi(1_1, 0_2)\phi(0_1, 1_2)\right]. \tag{12.4}$$

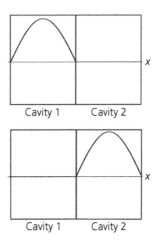

**Figure 12.4** *Top*, the wave function $\phi(1_1, 0_2)$ for the photon in cavity 1, with none in cavity 2. *Bottom*, the wave function $\phi(0_1, 1_2)$ for the photon in cavity 2, with none in cavity 1.

When we sum this over all the possible positions of the photon in the cavities, the first two factors, $\phi(1_1, 0_2)^2$ and $\phi(0_1, 1_2)^2$, are replaced by unity: if I know the photon is somewhere in cavity 1, then the probability of finding it anywhere in that cavity is unity; the same is true for cavity 2. On the other hand, in the last term, the two functions have no overlap, and their product vanishes. This just says that the cavities are in distinct places. Since the overlap term now vanishes due to the photon part, the probability at the screen becomes simply

$$P_{\text{at/ph}} = \frac{1}{2}\left(\psi_1^2 + \psi_2^2\right). \tag{12.5}$$

There is no interference term; the pattern on the screen becomes just a broad blur with no rapidly varying brightness bands. For each atom, we could have looked for the photon and identified the path, but even if we did not, the interference is gone. A graph of the possible positions of the atom on the screen might look like the one shown in Fig. 12.5.

But if we did not look, we can recover the interference pattern, that is, erase the above result in the following way. *After* each atom has gone through the cavities, suppose we open a photon detector at D in Fig. 12.3, allowing the cavities to see a photon detector. (The detector is

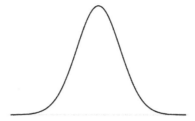

**Figure 12.5** Possible graph of number of atom counts versus position on the screen when there is no interference term. Of course, the individual atoms make dots on the screen that add up to this after many atoms have passed through the slits.

another single atom that can absorb the photon and go into an excited state). But if D absorbs the photon, we will no longer be able to look in the cavities to see which path the atom traversing the cavities took. We will have *erased* our previously available knowledge of path, and so we ought now to be able to see an interference pattern. Is this a way to change the past? After all, we can decide to open the detector or not, even after the atom has arrived at the screen. One author declared the experiment must be impossible, since he thought it did alter the past. Well, fortunately, it is not as weird as that!

Let's do a bit of algebra before we see how to do the erasing. When we open the shutter, the photon can be thought of as being in a different set of states; these are the symmetric and the antisymmetric states

$$\phi_+ = \frac{1}{\sqrt{2}} \left[ \phi(1_1, 0_2) + \phi(0_1, 1_2) \right], \text{ and} \qquad (12.6)$$

$$\phi_- = \frac{1}{\sqrt{2}} \left[ \phi(1_1, 0_2) - \phi(0_1, 1_2) \right], \qquad (12.7)$$

respectively. These new wave functions are shown in Fig. 12.6. Recall that, in finding the energy states in the double oscillation in Chap. 4, we found them to be similar symmetric and antisymmetric combinations of localized states. We have also done a similar "change of basis" before for spins in Sec. 5.3. We can rewrite the localized functions in terms of these symmetric and antisymmetric functions; just add and subtract the

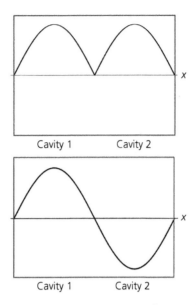

**Figure 12.6** *Top*, the symmetric wave function $\phi_+$ for the photon in the cavities. *Bottom*, the antisymmetric wave function $\phi_-$ for the photon.

two last equations:

$$\phi(1_1, 0_2) = \frac{1}{\sqrt{2}}(\phi_+ + \phi_-), \text{ and} \quad (12.8)$$

$$\phi(0_1, 1_2) = \frac{1}{\sqrt{2}}(\phi_+ - \phi_-). \quad (12.9)$$

If we substitute these into the atom–photon equation, Eq. (12.3), we get

$$\begin{aligned}
\Psi_{\text{atom-photon}} &= \frac{1}{\sqrt{2}}\left[\psi_1 \frac{1}{\sqrt{2}}(\phi_+ + \phi_-) + \psi_2 \frac{1}{\sqrt{2}}(\phi_+ - \phi_-)\right] \\
&= \frac{1}{\sqrt{2}}\left[\phi_+ \frac{1}{\sqrt{2}}(\psi_1 + \psi_2) + \phi_- \frac{1}{\sqrt{2}}(\psi_1 - \psi_2)\right] \\
&= \frac{1}{\sqrt{2}}(\phi_+\psi_+ + \phi_-\psi_-), \quad (12.10)
\end{aligned}$$

where we now have introduced the symmetric and the antisymmetric atom wave functions

$$\psi_+ = \frac{1}{\sqrt{2}}(\psi_1 + \psi_2), \qquad (12.11)$$

$$\psi_- = \frac{1}{\sqrt{2}}(\psi_1 - \psi_2), \qquad (12.12)$$

respectively. This math is important. Recall from our discussion of spin that the kind of wave functions we use depends on the experiment we will perform. In this case, the manipulations are necessary because it turns out that the detector D, another atom, will absorb a photon *only if it is in the symmetric state* $\phi_+$. So, we have two possibilities in Eq. (12.12) for the atom traveling through the cavities: (1) it comes out in state $\psi_+$, with D having absorbed the photon because the photon was in the necessary state $\phi_+$, or (2) the atom exits in state $\psi_-$ because the photon was in state $\phi_-$, and D did *not* absorb the photon. By looking at D, we can collapse the wave function to either of these two states. But if D did absorb the photon (DE, for D excited), then the atom heading for the screen has wave function $\psi_+$, and the pattern on the screen will be

$$P_{DE} = \psi_+^2 = \frac{1}{2}(\psi_1^2 + \psi_2^2 + 2\psi_1\psi_2). \qquad (12.13)$$

But if D did not absorb the photon (DU for D unexcited), the atom wave function is $\psi_-$, and then we have the probability

$$P_{DU} = \psi_-^2 = \frac{1}{2}(\psi_1^2 + \psi_2^2 - 2\psi_1\psi_2). \qquad (12.14)$$

These two patterns are shown in Fig. 12.7; the interference oscillations are restored in each case. But look closely, and you see that the two patterns are such that where the interference term is positive in the DE case, it is negative in the DU case, and vice versa, so that the peaks in one correspond to valleys in the other. Indeed, if you overlap them as in Fig. 12.8, you see that they would add up to exactly what resulted in Fig. 12.5.

Did we change the past? Not really. When we take data in the experiment, we must keep the two kinds of atoms hitting the screen separate if we are to see the interference: we might put a red mark at a screen site corresponding to a DE atom (for which D was excited) and put a green mark on the screen for a DU atom. When we did not use the detector at D, we had no way of separately marking the atoms, and they

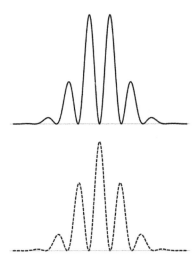

**Figure 12.7** *Top*, the interference pattern when the detector D does not absorb a photon. *Bottom*, the interference pattern when the detector D does absorb the photon. Note the one has peaks where the other has valleys, and vice versa.

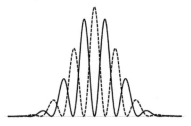

**Figure 12.8** A superposition of the interference patterns formed by DE (dashed line) and DU atoms (solid line), respectively. If added together, these would give the plot in Fig. 12.5.

were mixed up into one total screen blur. In that case, we could go back and see which path each took, depending on which cavity contained the photon for each atom. But, with D acting, we could separate the atoms into two kinds: those with DE, and those with DU; each will show interference when only the atoms of the same kind are taken into account. So, even if we made the decision to open D after the atom had struck the screen, we would still be able to post select whether to mark the atom position with red or blue (interference modes) or to leave it in the uncolored (which-way) mode. We are not changing the past, but it is still pretty weird.

## 12.2 Using photons and polarization

The eraser experiment has been done experimentally in several different ways, usually using polarized photons, which can be manipulated in rather exotic ways. Recall from Chap. 6 that the state of a photon polarized at 45° is a superposition of a horizontally polarized state and a vertically polarized state:

$$\psi_{45°} = \frac{1}{\sqrt{2}}(\psi_H + \psi_V). \tag{12.15}$$

In the same way, the state of a photon polarized at −45° (i.e., perpendicular to the one at 45°) is given by

$$\psi_{-45°} = \frac{1}{\sqrt{2}}(\psi_H - \psi_V). \tag{12.16}$$

The two polarization states are shown in Fig. 12.9.

We consider a two-slit experiment in which there is a source emitting photons (in a 45° state). These would go through a normal two-slit device and form an interference pattern on a screen. But suppose we put polarizers after each slit (or in front of them), as shown in Fig. 12.10. The interference pattern will disappear because states of perpendicular polarization do not overlap and so do not interfere with one another; the screen will have a sum of the patterns of each kind of photon, with no oscillation in the brightness of the pattern:

$$P_{\text{screen}} = \psi_{H1}^2 + \psi_{V2}^2, \tag{12.17}$$

where we have noted that the state corresponding to path 1 is H and the one corresponding to path 2 is V. We get no interference because we know the path the photon took. The pattern looks much like that in Fig. 12.5.

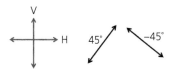

**Figure 12.9** The polarizations at 45° and −45° are superpositions of vertically (V) and horizontally (H) polarized states.

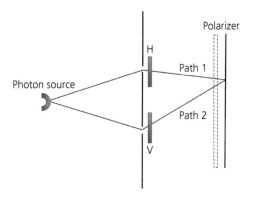

**Figure 12.10** A two-slit photon interference device. Polarizers are placed after the slits with a horizontal (H) polarizer for slit 1, and a vertical (V) polarizer for slit 2; these give which-way information. Another polarizer (dashed) can be placed in front of the screen; its orientation determines if one erases the which-way information and re-establishes interference patterns.

Suppose now we place a horizontal polarizer in front of the screen (see where the dashed rectangle is in Fig. 12.10). What that does is block out all the V state, so that we have now simply

$$P_{screen} = \psi_{H1}^2. \qquad (12.18)$$

There is still no interference pattern; we know the photon came by path 1. If we turn the polarizer so that it is vertical, we will block the other component and get just $\psi_{V2}^2$.

However, what happens if we rotate the screen polarizer to 45°? To see, we have to invert Eqs. (12.15) and (12.16) by adding them to, and subtracting them from, one another:

$$\psi_{H1} = \frac{1}{\sqrt{2}}\left[\psi_{45}^{(1)} + \psi_{-45}^{(1)}\right], \text{ and} \qquad (12.19)$$

$$\psi_{V2} = \frac{1}{\sqrt{2}}\left[\psi_{45}^{(2)} - \psi_{-45}^{(2)}\right]. \qquad (12.20)$$

The superscripts (1) and (2) on the states on the right indicate the path taken by the corresponding photons. But now the state at the screen—given that only 45° photons can get through—is

$$\Psi_{screen} = \frac{1}{\sqrt{2}}\left[\psi_{45}^{(1)} + \psi_{45}^{(2)}\right], \qquad (12.21)$$

with the probability

$$P^{(45)}_{\text{screen}} = \frac{1}{2}\left[\psi^{(1)2}_{45} + \psi^{(2)2}_{45} + 2\psi^{(1)}_{45}\psi^{(2)}_{45}\right]. \qquad (12.22)$$

We get an interference term again! The polarizer has removed our which-way knowledge of the photon, and so interference has returned. Similarly, if we rotate the polarizer at the screen by 90°, we allow only −45° polarization through, and we get

$$\Psi_{\text{screen}} = \frac{1}{\sqrt{2}}\left[\psi^{(1)}_{-45} - \psi^{(2)}_{-45}\right], \qquad (12.23)$$

with the probability

$$P^{(-45)}_{\text{screen}} = \frac{1}{2}\left[\psi^{(1)2}_{-45} + \psi^{(2)2}_{-45} - 2\psi^{(1)}_{-45}\psi^{(2)}_{-45}\right]. \qquad (12.24)$$

The interference term here has the minus sign, so the interference patterns in these two cases are related, just like those shown in Fig. 12.8. They are shifted relative to one another so they would give no interference if we did not distinguish between the two kinds of photons on the screen by use of the screen polarizer.

An article in the magazine *Scientific American* (May 2007) outlines how this experiment can supposedly be done by anyone with a laser pointer and a set of linear polarizers. Of course, the quantum effects are not evident when one uses a laser beam; the experiment can be explained by classical optics. But one can imagine doing it one photon at a time, so that quantum mechanics would be necessary for the explanation.

# 13

# Virtual Particles and the Four Forces

> Empty space is a boiling, bubbling brew of virtual particles that pop in and out of existence in a time scale so short that you can't even measure them.
>
> LAWRENCE M. KRAUSS

There are four known forces in nature: gravitational, electromagnetic, strong nuclear, and weak nuclear. A discussion of all the forces necessarily involves *every* phase of physics; this could be a very wide-ranging discussion. However, our goal has been to concentrate on the basic aspects of quantum mechanics, not to do a survey of *all* things quantum. So, we will limit our discussion here to showing that there is a very general common principle behind all the forces of nature, namely, the exchange of various kinds of "virtual" particles, which fits our search for weirdness in nature. One might think of the vacuum as something absent of everything, but it turns out to be very active; it seems to contain many fields, which are ever ready to produce particles of various kinds, both real ones and *virtual* ones. The latter are more ephemeral; they exist for just a flash and then disappear, even violating energy conservation for a brief moment, something that is allowed because of an energy–time uncertainty relation.

Discussing the forces will get us into a few side discussions having to do with the implications of these forces.

## 13.1 Survey of the four forces

Gravity is certainly familiar to everyone; the fact that we are pulled toward the earth is encoded in *Newton's law of gravitation*, which says that two bodies, 1 and 2, attract each other by a force that is proportional to the product of their masses, $m_1$ and $m_2$, and inversely proportional to the square of the distance $r$ between them. So, the force is as follows:

# Survey of the four forces

$$F_{\text{grav}} = G\frac{m_1 m_2}{r^2}, \qquad (13.1)$$

where $G$ is Newton's gravitational constant. Einstein's general theory of relativity explained gravitational force as being due to the distortion of space-time in the neighborhood of a mass. It is a bit like a heavy ball being on a mattress and denting is so that a neighboring heavy ball is induced to move toward the first ball by the distortion. If the balls are too far apart, they will not feel the distortion and will not roll together. However, the gravitational force, in principle, is never really zero at any distance and so is said to have an infinite range.

Electrical forces are almost as frequently evident as those due to gravity. When brushing your hair on a very dry winter day makes it fly up, it is because you have separated negatively charged electrons from the positively charged nuclei of their parent atoms. The positive charge of the nucleus normally attracts and binds the negatively charged electrons, and the net result is neutral, but when one separates the two and accumulates an excess of, say, electrons, in what is called a *static* charge, the effects are noticeable. Electrons and charged ions (nuclei with too few or too many electrons) get separated in clouds in a thunderstorm, and the charge can pass to or from the ground in the form of lightning. Of course every electrical object we have in our homes involves charges in motion (electron currents in lights, motors, etc.).

*Coulomb's law* for the attraction or repulsion of two electric charges is quite similar in form to the law of gravitation:

$$F_{\text{elec}} = k_e \frac{q_1 q_2}{r^2}, \qquad (13.2)$$

where the charges are $q_1$ and $q_2$, and $k_e$ is a constant of nature. Electric charge is measured in "Coulombs," which is a rather large unit: 1 Coulomb is the charge on $6.24 \times 10^{18}$ electrons; that's a lot of electrons! Charges can come in both positive and negative forms, with the proton having a positive charge ($+e$), and the electron, a negative charge ($-e$). If one of $q_1$ and $q_2$ is negative, the force is attractive; if both are positive or both negative, $F_{\text{elec}}$ is repulsive. Gravity has only an attractive force. The hydrogen atom, made up of a proton and an electron, is held together by the attractive Coulomb (electromagnetic) force. Electromagnetic interactions are so named because they also include magnetism, which arises from charges in motion. The electromagnetic force, like gravity, can act over large distances, since it falls off with $1/r^2$.

Unlike gravity, positive and negative charges can cancel each other out. If a positive charge has an equal uniform amount of negative charge around it, another charge will not see it.

The *strong nuclear force* is what holds the nucleus of the atom together. It acts on a short scale requiring neutrons and protons to be relatively near one another to feel the attraction. In fact, the protons and neutrons themselves are made up of quarks, and it is these particles that interact by the fundamental strong nuclear force. The forces between the quarks behave rather strangely, and we will return to their discussion in Sec. 13.6. Squishing several protons together in a nucleus means the nuclear force has to fight against the repulsion of the electrical force. The neutrons in the nucleus have no electrical charge and so mediate the proton electrical repulsion by separating the protons and by their own nuclear attractions to the other nucleons.

Isotopes are nuclei with the same proton number (and similar chemical properties) but differing numbers or neutrons. We have seen this before in $^3$He and $^4$He. Large nuclei and those with too many protons or too many neutrons can be unstable and are *radioactive*, that is, they can emit particles, which often have very large energies. The most common forms of radioactivity are rather unimaginatively called $\alpha$, $\beta$, and $\gamma$. In $\alpha$-radioactivity, an $\alpha$-particle ($^4$He nucleus) escapes the nucleus, changing it into a different element with two fewer protons and two fewer neutrons. (This form of radioactivity may be the best example of quantum tunneling; the $\alpha$-particle tunnels through the potential barrier formed by the combination of the short-ranged attractive nuclear force among the nucleons, and the long-ranged electrical repulsion between the protons.) The rate of decay of radioactive elements is measured by the half-life, which is the time it takes for half of the nucleons to decay. In $\gamma$-radioactivity, a high-energy photon is emitted from an excited nucleus; this process results from electromagnetic interactions. Nuclear *fission* occurs when a large nucleus, like that for uranium, breaks into two nearly equal-sized smaller nuclei, releasing the energy pent up in the proton–proton electrical repulsions. Nuclear reactors use this process as an energy source.

The *weak nuclear force* is responsible for $\beta$-*decay,* which occurs when a neutron turns into a proton, an electron, and an antineutrino. (The electron is the $\beta$-particle.) A neutron left on its own, that is, not bound in the nucleus of an atom, has a half-life of about 15 minutes. Neutrons confined within most atomic nuclei do not decay because it is energetically favorable for them to maintain stability. However, β-decay does occur naturally for elements with nuclei having an excess of neutrons. Carbon-12, with six protons and six neutrons (as well as six electrons) is the most abundant carbon isotope; carbon-14, with six protons and eight neutrons, has a half-life of 5730 years and β-decays into stable nitrogen-14, with seven protons. (This property makes it ideal for use in determining the age of ancient materials of organic origin—a technique called *carbon dating*.)

Neutrinos, almost massless particles, are emitted in β-decay and interact with matter via the weak force. A giant detector called Ice Cube, buried in the ice in Antarctica, is being used to do neutrino astronomy by detecting neutrinos from violent astrophysical sources. The most energetic of them get individual names, like Bert or Ernie. The weak force is a vital element in the fusion processes that takes place in the sun, as we will discuss in Sec. 13.2.

## 13.2 Sizes of the forces

It is useful to compare the sizes of these forces with one another. The names "weak" and "strong" are relative adjectives, as we will see. If we divide $F_{\text{grav}}$ by $F_{\text{elect}}$, the dependence on distance $r$ cancels out, and if we consider two protons, then this ratio is

$$\frac{F_{\text{grav}}}{F_{\text{elec}}} = \frac{Gm_p^2}{k_e e^2} = 2 \times 10^{-37}, \qquad (13.3)$$

where $m_p$ is the proton mass, and $e$ is the proton charge. The gravitational force is attractive, and the electric force is repulsive, so this is just the ratio of the strengths. Gravity is *way* weaker than the electric force! Nevertheless, we are so much more aware of gravity than of electrical forces in everyday life because the latter's positive and negative charges can cancel each other's effects.

There is a way of writing the strength of a force in terms of a pure number called the "coupling constant." The quantity $Fr^2$ has units of force times distance squared (energy times distance), but those are also the units of the product $\hbar c$, where $\hbar$ is Planck's constant divided by $2\pi$, and $c$ is the velocity of light. Thus, $Fr^2/\hbar c$ is unitless. Indeed,

$$\alpha = \frac{F_{\text{elect}} r^2}{\hbar c} = \frac{k_e e^2}{\hbar c} = \frac{1}{137.04} = 0.0073 \qquad (13.4)$$

is called the *fine structure constant* in atomic physics and is a measure of the strength of the electromagnetic interaction. The fact that it is a small number indicates that the electromagnetic interaction is, in some sense, weak itself. The fine structure constant is one of the most accurately known constants of nature, with an accepted value of 0.0072973525698. Knowing something to ten digits is equivalent to knowing your height to the radius of one atom. The electron's so-called g-factor, having to do with the electron's magnetic properties, is known to 13 significant figures!

How does the strong nuclear force compare to the electric force? That is harder to quantify, because the ranges of the two are so different. The nuclear force has a short range, being sizable between nucleons for distances only of $10^{-15}$ m, while the electric force drops off much more slowly. But we can compare them when both are acting within a nucleus. Helium has a nucleus that is about $2 \times 10^{-15}$ m in radius. So, suppose we consider the relative size of the nuclear force and the electric force at such a distance. The nuclear force has $F_{nuc} r^2/\hbar c \sim 1$ at this range, making it about 137 times bigger than the electromagnetic force when two protons are very close, so this force can bind a nucleus together quite tightly, despite all the protons in it with their mutual electric repulsion. However, when nuclei get very large, as for uranium or plutonium, the electric force can win out when nuclear fission happens, breaking a large nucleus into smaller fragments.

The coupling constant for the weak nuclear force, the ratio $F_{weak} r^2/\hbar c$, is about $10^{-6}$, which makes it about million times weaker than the strong nuclear force. It operates over a very short distance, with a range of about $10^{-18}$ m.

We will return to a more detailed discussion of these fundamental forces in Sec. 13.9.

## 13.3 Virtual particles

The source of any interaction in nature is the exchange of *virtual* bosons between the interacting particles. The vacuum is a very busy place; particles that were not there suddenly appear. For example, a γ-ray, which is a form of high-energy electromagnetic radiation, like visible light but at higher frequency and energy, can turn into an electron and anti-electron (positron) pair if it has sufficient energy (and a nearby nucleus or other particle to help conserve momentum). The general process in which a γ-ray converts into a particle–antiparticle pair is called *pair production*. Alternatively, when an electron meets a positron, the two particles will destroy each other to form a γ-ray in *pair annihilation*. A free neutron, with a half-life of 15 minutes, will decay into a proton, an electron, and an antineutrino. The theory of these processes assumes that the vacuum contains a "field" for each of these particles, whether the field is activated with the actual existence of a particle or not. The neutron and proton are made up of quarks, so it is a quark field that has a weak interaction and changes in the neutron decay.

We need to take a mathematical diversion to help understand virtual particles. It uses only algebra and is worth the effort because it gives quite a bit of insight. In fact, in this analysis, we see a bit of one of the ways physics uses mathematics. Often in solving, say, the Schrödinger equation, we find a formula supposedly describing our experiment; but it is only in "taking apart" the mathematics by some further analysis in which we devise words or drawings describing the mathematics that we figure out what it is trying to tell us; that is, what the "real" physics is in the processes. This occurs here when we describe the mathematical formulas by so-called Feynman diagrams.

In our mathematical model, we will consider an atom in an *excited state* designated by $|es\rangle$. We will assume the atom has only one other quantum state: a lowest state, or *ground state*, designated by $|gs\rangle$. In the excited state, the atom can emit a photon and fall to the ground state, or the atom in the ground state might absorb an already existing photon and move to the excited state. This two-level model is realized in actual experiments, often using the same alkali atoms that form Bose–Einstein condensates. Suppose the atom is in a *cavity*, which is a conducting box that gives boundaries to form standing quantum waves for the radiation. Only certain allowed frequencies of radiation, quantum states, can persist in the cavity. We assume the atom emits or absorbs only one possible photon of frequency, $\omega_p$. Our single atom has an excited state with energy

$$E_1 = E_{es}. \tag{13.5}$$

The other possible state of this system then is the atom in its ground state $E_{gs}$, but now with one photon present, so the system energy is

$$E_2 = E_{gs} + \hbar\omega_p \tag{13.6}$$

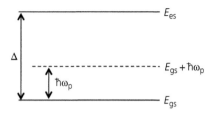

**Figure 13.1** Atom and photon energies. The excited atom state $E_{es}$ energy is larger than the ground state energy $E_{gs}$ by $\Delta$. The cavity system of atom and photon has the energy states $E_1 = E_{es}$, and $E_2 = E_{gs} + \hbar\omega_p$. The photon energy is shown with $\hbar\omega_p < \Delta$, but we can also arrange to have $\hbar\omega_p > \Delta$.

(see the energy level diagram in Fig. 13.1). The atom excited-state energy $E_{gs}$ is larger than the ground state energy $E_{es}$ by $\Delta$; the photon energy can be smaller or larger than $\Delta$, although we show it smaller in the figure. The energies $E_1$ and $E_2$ are the system energies when the photon and atom do not interact. The corresponding wave functions are $\psi_1 = |es, 0\rangle$, where the $es$ represents the atom excited state, and the 0 represents the vacuum state of the photon, that is, zero photons, and $\psi_2 = |gs, 1\rangle$, where the atom is in its lowest state and there is one photon present. When we turn on the interaction between the atom and the photon, the atom can then emit or absorb a photon, and the *wave functions and energies will be changed* because of the interaction. Indeed, now we will find that the new wave functions will each be a superposition of the old ones:

$$\psi = A|es, 0\rangle + B|gs, 1\rangle, \quad (13.7)$$

with two different sets of $A$ and $B$, each corresponding to how one of the original two noninteracting states evolves when the interaction is turned on. The energies will be altered by one increasing in energy, and the other decreasing by the same amount, as we will see.

It turns out that quantum theory shows that the interaction between an atom and a photon results in new energies $E$ being the solutions of a simple quadratic equation for our simple two-level system. We would need to learn more details about the Schrödinger equation to derive this equation, but if we accept it, we can get some insight into the nature of virtual particles. The energy equation is

$$E^2 - (E_2 + E_1)E + E_1 E_2 - Q^2 = 0. \quad (13.8)$$

In this, $Q$ is the coupling energy between the atom and the radiation. It is related to the fine structure constant $\alpha$ that we saw before for the electromagnetic force coupling two charged particles. Classically, an electron will radiate if it accelerates. Electrons in a cell phone tower are oscillating in the antenna and giving off microwave radiation. Cell phones have very small antennas to pick up the microwave radiation and to radiate their own signals. Quantum mechanically, when an electron in an atom moves from one quantum level to a lower one, it is accelerating and so emits a photon. Equation (13.8) above is the standard quadratic form, with $ax^2 + bx + c = 0$ having solutions $x = (1/2a)(-b \pm \sqrt{b^2 - 4ac})$ or, here,

$$E = \frac{1}{2}\left[(E_2 + E_1) \pm \sqrt{(E_2 + E_1)^2 - 4(E_1 E_2 - Q^2)}\right]. \qquad (13.9)$$

There are two solutions: one with the plus sign, and the other with the minus sign. Multiply out the square inside the square root and combine with $-4E_1 E_2$ to get $(E_1 + E_2)^2 - 4E_1 E_2 = (E_1 - E_2)^2$, so

$$E = \frac{1}{2}\left[(E_1 + E_2) \pm \sqrt{(E_1 - E_2)^2 + 4Q^2}\right]. \qquad (13.10)$$

The usual situation is that the interaction energy $Q$ is much smaller than the energy difference $E_1 - E_2$. In this case, we can make an approximation. First, factor out $(E_2 - E_1)^2$ from the square root to give an exact equality:

$$E = \frac{1}{2}\left[(E_1 + E_2) \pm (E_1 - E_2)\sqrt{1 + \frac{4Q^2}{(E_1 - E_2)^2}}\right]. \qquad (13.11)$$

Now, we turn this into what is called an *infinite series*. The geometric series is a well-known example of this kind of thing; perhaps you recognize this sum:

$$S_n = \sum_{k=0}^{n} x^k = 1 + x + x^2 + \cdots + x^n = \frac{1 - x^{n+1}}{1 - x}. \qquad (13.12)$$

It is fairly easy to prove the equality of the last two forms. Simply multiply both sides of equation by $(1 - x)$, and then multiply out the terms;

all but the first and last terms of the middle piece cancel, confirming the equality with the right side. But now if $n \to \infty$, and $x$ is between $-1$ and $1$, then $x^{n+1} \to 0$ as $n$ becomes large, and we have

$$S_\infty = \frac{1}{1-x} = 1 + x + x^2 + \cdots, \qquad (13.13)$$

where the series has an *infinite* number of powers of $x$. It is interesting to see how many terms are needed to get a certain accuracy. For example, suppose $x = 0.01$; then, the exact result is $1/0.98 = 1.0204\ldots$. The sum of the first two terms of the series, $1 + x$, is equal to $1.02$. The sum of the first three terms gives $1.0204$. The series converges quickly when $x$ is small. But if $x = 0.98$, the exact result is $50$, but the sum of the first four terms gives $2.8828$—not accurate at all. One needs over $450$ terms to get two-decimal-place accuracy. But the series still works.

For our case of Eq. (13.11) we have the form $\sqrt{1+x}$, which also has an infinite series:

$$S = \sqrt{1+x} = 1 + \frac{1}{2}x - \frac{1}{8}x^2 + \frac{1}{16}x^3 - \cdots. \qquad (13.14)$$

If $x$ is very small, then we might get away with considering only the first two terms in the sum as a good approximation. In our case, we have

$$x = \frac{4Q^2}{(E_1 - E_2)^2}. \qquad (13.15)$$

We take this to be very small because $Q$ is small compared to $E_1 - E_2$. Then, keeping only $1 + \frac{1}{2}x$ in Eq. (13.14) gives

$$E \approx \frac{1}{2}\left\{(E_1 + E_2) \pm (E_1 - E_2)\left[1 + \frac{2Q^2}{(E_1 - E_2)^2}\right]\right\}, \qquad (13.16)$$

where $\approx$ means "approximately equal to." Multiply this out, and we get our two solutions, one with the plus sign and one with the minus sign:

$$E \approx \frac{1}{2}\left[(E_1 + E_2) \pm (E_1 - E_2) \pm \frac{2Q^2}{(E_1 - E_2)}\right] \qquad (13.17)$$

$$= \begin{cases} E_1 + \frac{Q^2}{(E_1 - E_2)} \\ E_2 - \frac{Q^2}{(E_1 - E_2)} \end{cases}$$

Virtual particles 149

or, putting in the values of $E_1$ and $E_2$ from Eqs. (13.5) and (13.6), we have

$$E_1 - E_2 = E_{es} - E_{gs} - \hbar\omega_p = \Delta - \hbar\omega_p. \qquad (13.18)$$

The new energies are

$$E_+ = E_{es} + \frac{Q^2}{(\Delta - \hbar\omega_p)}, \text{ and} \qquad (13.19)$$

$$E_- = E_{gs} + \hbar\omega_p - \frac{Q^2}{(\Delta - \hbar\omega_p)}. \qquad (13.20)$$

The state $E_1 = E_{es}$ is changed to $E_+$ by the small correction due to interaction with the photon field. It can be positive or negative, depending on the sign of $\Delta - \hbar\omega_p$. If the latter is positive, then the new state has higher energy, and the other state, $E_2 = E_{gs} + \hbar_{es}$, is lowered to $E_-$ by an equal amount.

The Schrödinger equation also gives the solution for the new wave functions. We find the new states corresponding respectively to these two energies, $E_+$ and $E_-$, to be given by

$$\psi_+ = C\left[|es, 0\rangle + \frac{Q}{(\Delta - \hbar\omega_p)}|gs, 1\rangle\right], \text{ and} \qquad (13.21)$$

$$\psi_- = C\left[|gs, 1\rangle - \frac{Q}{(\Delta - \hbar\omega_p)}|es, 0\rangle\right]. \qquad (13.22)$$

The state with no photon, $|es, 0\rangle$, now has a bit of the state with one photon in it. The other energy state, $|gs, 1\rangle$, now has a bit of the state with no photon in it. ($C$ is a normalization constant, so the total probability comes out 1 in either state.)

Now, how do we interpret these results? We have mathematical answers, but the physics is not complete without words and pictures to interpret that math. In Fig. 13.2, we give a graphical interpretation for the mathematical form of the energy. Consider Fig. 13.2(a), which shows a diagram that corresponds to the energy $E_+$. In this case, we are seeing a correction to the energy of excited state $E_{es}$. In the diagram, this is called the "initial" state, where we are tracking processes supposed to be happening in time. We will write the correction to the energy in the

# Virtual Particles and the Four Forces

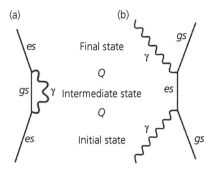

**Figure 13.2** Feynman diagrams for one atom interacting with the radiation field. (a) The initial excited state $|es, 0\rangle$ emits a virtual photon $\gamma$ and moves into the intermediate $|gs, 1\rangle$ that is, the ground state plus a photon. In a very short time, the photon is reabsorbed, and the atom goes to the final state, which is again the original state. (b) The state of one atom in the initial $|gs, 1\rangle$ goes to an intermediate state, $|es, 0\rangle$, with the photon absorbed, but it soon returns to the original state by reabsorbing the photon.

stretched-out form

$$Q \frac{1}{(E_{es} - E_{gs} - \hbar\omega_p)} Q = \frac{Q^2}{(\Delta - \hbar\omega_p)}. \tag{13.23}$$

There is first an interaction of strength $Q$; this interaction corresponds to the emission of a photon, so the atom goes to the "intermediate state," that is, into $E_{gs} + \hbar\omega_p$, which is represented by the denominator having the energy difference between that state and the initial state $E_{es}$; then, there is another $Q$, for the atom's reabsorption of the photon to lead back to the final state, which is the original one again. Figure 13.2(b) interprets the change in energy of the state $|gs, 1\rangle$ in the same diagrammatic way. This initial state of energy $E_{gs} + \hbar\omega_p$ then gives

$$Q \frac{1}{(E_{gs} + \hbar\omega_p - E_{es})} Q = -\frac{Q}{(\Delta - \hbar\omega_p)}, \tag{13.24}$$

with $Q$ at the absorption of the photon to get to the intermediate state $E_{es}$, with the energy difference with the initial state given in the denominator; then, the photon is reemitted with interaction $Q$. The resulting graphs are called *Feynman diagrams*. Richard Feynman won the Nobel Prize in 1965 for the development of *quantum electrodynamics,* the study of charges interacting with radiation; in order to help classify and keep track of the many mathematical terms that one must include in the theory, one draws such diagrams.

The math says interactions with the electromagnetic field cause an atom in the excited state to occasionally jump into the ground state by emitting a photon. But this process *does not conserve energy*: $E_{es}$ is not equal to $E_{gs} + \hbar\omega_p$, that is, $\Delta - \hbar\omega_p \neq 0$, and so the atom reabsorbs the photon almost immediately! Such short-existence photons are *virtual* particles. The photon manages to get away with this violation for a short time because of the uncertainty principle for energy, which says that the uncertainty in energy $\Delta E$, and that in time $\Delta t$, satisfy $\Delta E \Delta t \geq \hbar$. So, as long as the time of existence of the photon is short enough, then energy conservation can seem to be violated.

We neglected terms in $x^2$ and higher order in our series expansion in Eq. (13.16). These next orders involve the photon being emitted a second time, and a third time, etc., in the one case or absorbed many times in the other case. These situations correspond to more complicated Feynman diagrams. The one in $x^2$ will have $Q^4$ in the energy, corresponding to two emissions and two absorptions, in any diagram. The original result, Eq. (13.10), before we expanded it in $x$, was exact and summed up *all* of these possible terms to all orders in $Q$! But, by expanding it, we have learned what the various pieces mean physically. Making the formula more complicated actually simplified its meaning.

We have assumed that the two initial energies $E_1$ and $E_2$ were unequal, that is, that $\Delta - \hbar\omega_p \neq 0$. But suppose that the photon energy matches the separation between the two atom states, that is, $\Delta = \hbar\omega_p$ in Fig. 13.1. Then, our expansion technique is no longer valid because $x$ of Eq. (13.15) is infinite! If we look back at this case in our *exact* energy solution, Eq. (13.10), we see that the equality of the two energies makes the equation simplify to

$$E = \frac{1}{2}\left[(E_1 + E_2) \pm \sqrt{(E_2 - E_1)^2 + 4Q^2}\right] = E_e \pm Q \qquad (13.25)$$

when we put $E_1 = E_2 = E_e$. There is only one power of $Q$ in this now because the photon is emitted and *not* reabsorbed; it just goes on independently (or, in the other case, the photon that was already present is absorbed and not reemitted). A *real* photon can be created here, which is allowed because energy is conserved in the process and what was the intermediate state can continue to exist indefinitely. The wave functions corresponding to the $\pm$ energies given are

$$\psi_{\pm} = \frac{1}{\sqrt{2}} \left( |es, 0\rangle \pm |gs, 1\rangle \right). \tag{13.26}$$

Each state is a complete superposition of the two original states in equal amounts. The math has changed dramatically, and so has the physics. If one makes a measurement on this superposition, one will find one element or the other, of course, as in the case of any superposition. But the superposition of the two states allows a lowering of the energy of one of the states by $Q$, and a corresponding raising of the energy of the other state. If you start the system out in one of the states, for example, the excited atom state, it will evolve into the other one, the ground state plus photon, and then back into the first, and so on (i.e., if the atom and the photon are in the cavity, and the photon cannot escape to infinitely far away when it is emitted, it will ultimately be reabsorbed—but not necessarily immediately). This is reminiscent of the case in Chap. 4, where a single particle was in a double well. The particle could tunnel into the other well so that the correct wave functions were the sum and difference of the right-well state and the left-well state. If you started the particle on the right, it would tunnel to the left and back again. The periodic transitions of an atom between excited and ground states in this way are called "Rabi oscillations," named for Isidor Rabi, who won the 1944 Nobel Prize for his work in developing NMR, a basic part of present-day MRI.

In this analysis, only one particle was present, and any virtual photon emitted was reabsorbed by the same particle. What if there is a second particle to absorb the photon emitted by the first? Let's start out with *two atoms* present: one in an excited state, and one in the ground state, with no photon initially present. This state is $|es, gs, 0\rangle$; the corresponding energy is $E_1 = E_{es} + E_{gs}$. The second possible state is $|gs, gs, 1\rangle$, two ground state atoms with a photon, with energy $E_2 = 2E_{gs} + \hbar\omega_p$. When we turn on the interaction between the atom and the radiation field, we get energies that satisfy a quadratic again:

$$\begin{aligned} E &= \frac{1}{2}\left[ (E_2 + E_1) \pm \sqrt{(E_2 + E_1)^2 - 4(E_1 E_2 - 2Q^2)} \right] \\ &= \frac{1}{2}\left[ (E_1 + E_2) \pm (E_1 - E_2)\sqrt{1 + \frac{8Q^2}{(E_1 - E_2)^2}} \right] \\ &\approx \frac{1}{2}\left[ (E_1 + E_2) \pm (E_1 - E_2) \pm \frac{4Q^2}{(E_1 - E_2)} \right]. \end{aligned} \tag{13.27}$$

# Virtual particles

We have done an expansion, as in Eq. (13.14), and kept just the first terms, with $x = 8Q^2/(E_2 - E_1)^2$. The only difference here from the corresponding case of Eq. (13.16) is a factor of 4, instead of 2, appearing in the last term. The energy difference in the denominator is now $E_1 - E_2 = E_{es} + E_{gs} - 2E_{gs} - \hbar\omega_p = E_{es} - E_{gs} - \hbar\omega_p = \Delta - \hbar\omega_p$, just as in the previous case, leading to

$$E_+ = E_{es} + E_{gs} + \frac{2Q^2}{(\Delta - \hbar\omega_p)}, \text{ and} \quad (13.28)$$

$$E_- = 2E_{gs} + \hbar\omega_p - \frac{2Q^2}{(\Delta - \hbar\omega_p)}. \quad (13.29)$$

Things are changed from the one-atom equation by having an extra 2 in front of $Q^2$ in the last forms. Consider the final form $E_-$ first. Here, the initial state has a photon and two ground state particles. The intermediate state is the same as in Fig. 13.2(b), where one of the particles, say A, absorbs the photon and then reemits it while particle B continues on. But particle B could have also done the same thing. So, the result is a superposition of both possibilities, and the energy correction now has a 2 in front of the correction, corresponding to both atoms' corrections.

The result in the final form for $E_+$ is more interesting. As before, there is a contribution from the excited atom emitting and then quickly reabsorbing the photon. This is just the process shown in Fig. 13.2(a), and we get the same energy correction for that process. But the other half of the energy correction involves a new process where the emitted photon is reabsorbed by the *other* ground state atom, as seen in Fig. 13.3. The two atoms have *exchanged* the photon. If we have $\hbar\omega_p > \Delta$, then the

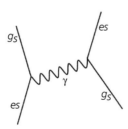

**Figure 13.3** Feynman diagrams for two atoms interacting with the radiation field. The excited state *es* emits a virtual photon γ and moves into the ground state *gs*. The ground state atom absorbs the photon. The pair energy is changed.

photon exchange will lead to an energy lowering due to the process, corresponding to a force of attraction. Here, our simple model does not give a distance dependence for this energy and the corresponding force. To get such a complete description of the electromagnetic interaction would require the exchange of photons of all possible wavelengths in free space. But the important feature here is that the energy change is due to *virtual* photon exchange; the intermediate state has an energetic photon in it that does not quite conserve energy. If the energy denominator in the expression were to vanish ($\Delta - \hbar\omega_p = 0$), then we could not have made our expansion of the square root; that would correspond to a real photon emission and not a virtual one. It is such an exchange of virtual particles that is the source of all the fundamental forces. *In every case, the particle exchanged is a boson, causing forces between fermions.* Let's look at each in detail.

## 13.4 Photons and the electromagnetic force

Here, we replace our simple two-state atom by an electron in free space, with the electron having any momentum and energy, and, similarly, our single photon by a photon having arbitrary energy and momentum. There are single-particle processes like those shown in Fig. 13.2. An electron can emit and reabsorb virtual photons of any momentum, and we have to sum over all the possible virtual momentum in diagrams like that in the left-hand side of Fig. 13.4. These processes are called *self-energy* processes. The photon itself has self-energy processes; recall that

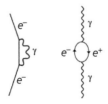

**Figure 13.4** Feynman self-energy diagrams. *Left*, the simplest diagram for an electron emitting and reabsorbing a photon represented by $\gamma$. Our simple mathematical model had such a diagram. *Right*, a photon creating a virtual electron–positron pair. The electron $e^-$ on the left has the forward-directed arrowhead; the positron $e^+$ on the right has the backward-arrowhead because it can be thought to be an electron traveling "backward" in time.

we have mentioned that the photon can create a real electron–positron pair if it has enough energy (and there is another particle nearby to allow momentum conservation). The lowest order self-energy effects are shown in the right-hand side of Fig. 13.4. Higher orders involving many photons or many electron–positron pairs are possible. The photon process is known as *vacuum polarization* because the vacuum has been separated into positive and negative charged parts, that is, it has become polarized. When we calculate the self-energies, we sum over all virtual momenta, that is, to arbitrarily high momenta. Processes like these contributing to the electron's *self-energy* give an infinite energy result! Since mass and energy are equivalent, it seems the electron has an infinite mass. A similar problem occurs in classical physics. Suppose the electron is a uniform sphere of charge; then, each small portion has an electromagnetic energy of interaction with every other small portion of charge. We can sum over all these interactions and get a value. But there is strong evidence that the electron is a point particle, and the classical self-energy then diverges (becomes infinite) as the charged pieces become closer together. The quantum theory calculation gives a divergent answer too. We know the actual electron mass is finite; one possible explanation is that our quantum theory is not valid at extremely high energies and momenta. In any case, there are ways of subtracting off the infinite parts of our answers in quantum electrodynamics and getting finite results. The process is called *renormalization*. What is left over after subtracting off the infinities is the real electron mass. This seemingly questionable process has been greatly refined in recent years so that it no longer causes nightmares to particle physicists.

The self-energy processes also affect the photon exchange processes that give rise to the electromagnetic force. How this might happen is shown in Fig. 13.5, where we consider the interaction between a proton and an electron, with some self-energy terms included. When one includes the electron self-energy into the calculation of the electron states of the hydrogen atom, there is a slight shift in the energy differences between certain states because the state where the electron comes closest to the proton is changed more in energy than the other state. The effect is proportional to the fine structure constant to the fifth power ($\alpha^5 = 2 \times 10^{-11}$), compared to the normal atomic energy proportional to $\alpha^2 = 1 \times 10^{-4}$. This *Lamb shift* is very small but was experimentally observed by Willis Lamb in 1947 and explained by Hans Bethe in the same year.

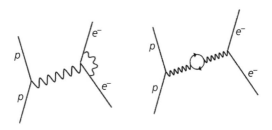

**Figure 13.5** Feynman diagrams for particle–particle interactions. *Left*, a correction term to the simplest exchange interaction. Here, the electron emits a second photon before absorbing the first and then reabsorbing the second. *Right*: A virtual exchanged photon creates a virtual electron–positron pair during exchange.

The interactions between photons and electrons can lead to very complicated quantities to be calculated. For example, consider the diagram in Fig. 13.6. In this example, we see that the interaction occurring when an electron absorbs a single photon can involve lots of activity. The strength of the emission or absorption of a photon depends on the size of the electric charge $e$. But what we observe experimentally must include all diagrams such as this. These computations also lead to infinity, and so the real charge must include all such diagrams renormalized to remove the divergences.

The electromagnetic force depends on the charges on the two particles, with the elementary charge being designated by $e$. Forces are

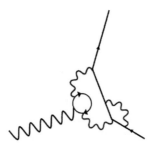

**Figure 13.6** A Feynman diagram for the photon absorption "interaction vertex." This is one possible diagram showing a renormalization of the bare interaction by electron–positron pair formation as well as photon exchanges. An infinite number of such diagrams are possible.

always caused by the exchange of a boson; in this case, it is the photon, as we have seen. What determines the range of a force is the mass of the exchanged particle; since the photon has zero mass, the range is infinite, dropping off only as $1/r^2$. Where the exchange boson has mass, the force decreases exponentially with distance, with the rate faster the larger the mass, as we will see below. Careful experiments to test the rate of distance dependence have always confirmed that the photon mass vanishes.

Two more comments about vacuum polarization. First, if the vacuum around a proton is constantly being split into virtual positrons and electrons, those electrons will be more attracted to the proton, and the positrons repelled. Thus, the proton's charge is shielded somewhat by the virtual electronic charge cloud surrounding it. That means the effective proton charge is smaller when seen from far away than from very close. This increase in charge when close is seen in proton–proton collisions in high-energy accelerators.

The gravity around a black hole is strong enough to "pull the vacuum apart" into an electron–positron pair. But suppose the black hole swallows the positron, but the electron is created sufficiently far from the hole to escape from the gravity of the black hole. Energy conservation requires that the particle falling into the hole have negative energy, which subtracts from the black hole mass. Basically, what has happened is that the black hole has emitted a particle. This is called *Hawking radiation*, after Stephen Hawking. Small black holes can disappear, in principle, from emitting this radiation.

An effect proposed by Hendrik Casimir in 1947 dramatically demonstrates the reality of the active vacuum we have described. Casmir's idea was to place two metal plates close to one another; because the space is restricted between the plates, the wavelength of any photon in that region is also restricted to be equal to or smaller than the distance between the plates. Larger wavelengths don't fit, just as in Fig. 1.3, which shows waves on a string. Any virtual photons emitted between the plates are restricted in wavelength, while those outside the space between the plates are not. This implies that the vacuum energy is decreased by putting the two plates close to one another, so there is a force of attraction between the plates. This force varies as $L^{-4}$, where $L$ is the distance between the plates. The effect was verified quantitatively by Steve Lamoreaux in 1997. Virtual vacuum fluctuations are directly observable.

## 13.5 Gravitons

The exchange boson in the gravitational force is called the *graviton*. It has spin $2\hbar$ and is massless, so the force goes again as $1/r^2$. Of course, massive objects can emit real gravitons as well as exchange virtual ones. There has been a large experimental effort to observe gravitational radiation, even of the classical sort that does not require a quantum description. Gravitational observatories that use very large laser interferometers have been built or are under construction in several places in the world. Such observatories are known as LIGOs, for "Laser Interferometer Gravitational-Wave Observatory." The idea is that a gravity wave can stretch matter in one direction while shrinking it in the perpendicular direction. Thus, the two arms of the interferometer would change relative lengths, leading to a change in an interference pattern. Two widely separated interferometers should see the coincident changes that would verify that the source was gravitational radiation. The first version of these detectors saw no signals but, almost immediately after the improved detector went online in 2016, the LIGOs in Washington and Louisiana saw gravity waves from the collision and merging of two black holes.

Gravity waves had already been observed indirectly. In an experiment at University of Massachusetts Amherst, Joseph Taylor and Russell Hulse, the latter then a graduate student, discovered a binary pulsar. A pulsar is a collapsed neutron star (a star that has undergone a supernova explosion and collapsed under gravity until all the protons were converted by weak interactions into neutrons). Pulsars emit a beam of radiation and matter, with this beam rotating with the star like the beacon of a lighthouse. The pulsar these two researchers found had a very large companion star in orbit around it. With two massive objects so close together in orbit, deviations from Newton's laws of gravitation as predicted by Einstein's general theory of relativity should be observable. The pulsing of the neutron star is a very reliable clock so that timing of the orbit is possible. When the pulsar recedes from the earth, the pulses decrease in frequency; when it approaches, they are higher in frequency, and this gives information on the orbit of the binary. What was most interesting is that the rate of pulses diminished over time, showing that the system was losing energy. The rate of loss agrees precisely with that predicted by Einstein's theory for loss by gravitational radiation. Taylor and Hulse won the 1992 Nobel Prize for this work.

A quantum theory of gravity has as yet eluded theorists, although, as in the case of Hawking radiation, some progress has been made. Recently, theorists have been able to derive a quantum correction to the general relativistic formula for the bending of light due to the gravitational distortion of space.

## 13.6 Gluons and the strong interaction

The proton is made up of three *quarks*. A quark is a fermion with spin 1/2 and a charge of 1/3$e$ or 2/3$e$, where the elementary charge is $e$. There are six quarks, with corresponding antiquarks: *up, down, strange, charm, bottom,* and *top*. The quarks combine to make the *hadrons*, which are the heavy particles of nature, and for which there are two families: *baryons*, including the proton and the neutron, and *mesons*, which include pions, kayons, and others. The proton is composed of two up quarks, and a down quark, and the neutron has one up and two down quarks. The quark model of the hadrons was proposed by Murray Gell-Mann and George Zwieig, and also by Yuval Ne'eman, independently, in 1964. The name "quark" arises from a line in a poem by James Joyce's book *Finnegans Wake*: "Three quarks for Muster Mark!"; the "three" is the number needed for the baryons. The up quarks have a charge of +2/3$e$, while the down quark has −1/3$e$, giving the proton a net charge of +$e$, and the neutron zero charge.

The mesons require only two quarks. For example, the $\pi^+$ has one up and an anti-down, and the $\pi^-$ has one down and an anti-up, while the $\pi^0$ is a quantum superposition of up–antiup and down–antidown. There are quite a few other mesons: $\rho, \eta, \eta', \phi, \omega, Y, \theta, J/\psi$, K, B, D, and T. Observations of other exotic particles, including mesons with four or five quarks, have been reported. The present theory of elementary particles that includes the quarks is called the *Standard Model* and has been remarkably successful.

Hideki Yukawa suggested in 1935 that the strong nuclear force was mediated by the exchange of $\pi$ mesons, and this is still the accepted view; however, the Standard Model postulates a new particle, the "gluon," as being the exchange boson between the quarks; it is so named because it "glues" the quarks together within the nucleus. The pion corresponds to the exchange of a quark pair between different nucleons. Figure 13.7 shows Yukawa's original idea and the modern version of it. When the proton and neutron are replaced by quarks, we can see that the Yukawa pion exchange is replaced by the exchange of a pair of quarks arising from gluon exchange interactions.

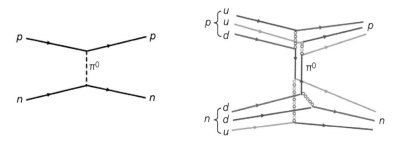

**Figure 13.7** Feynman diagrams for the strong nuclear interaction. (a) Yukawa's suggestion for the pion mediation of the nuclear force between a proton $p$ and a neutron $n$. (b) The same interaction in terms of quarks; $d$ indicates a down quark, and $u$ indicates an up quark. The bubble lines between quarks show the gluon exchange within the nuclei. The pion quarks are also exchanging gluons. (See http://en.wikipedia.org/wiki/Nuclear_force for an animation of the pion exchange in terms of quarks.)

So, we have a two-element nuclear force. The nuclear force between nucleons with the exchange of a pion is a result of an even stronger force between quarks mediated by exchange of gluons. The pion-exchange nuclear force is strongly attractive, with a short range of about $2 \times 10^{-15}$ m. It has short range because the particle being exchanged is massive (about 1/10 the mass of the proton). Contrast that with the long range of the electromagnetic force and that of gravity, where the particles exchanged are massless. The gluon exchange force has some strange properties. As the energy increases and the distance between quarks decreases, the gluon force becomes weaker, a phenomenon known as *asymptotic freedom*. On the other hand, when the distance between quarks increases, the force becomes stronger and indeed is so strong that free quarks are never seen! They must always combine in at least pairs. In measurements of electrical charge, one always sees $\pm e$, and not 1/3 or 2/3 of that. When a quark is exchanged between nucleons, it is always paired up with another quark to form the $\pi$ meson.

## 13.7 The $W$ and $Z$ mesons and the weak interaction

The weak nuclear interaction is mediated by a set of massive mesons, the $W^+$ (positively charged), $W^-$ (negatively charged), and $Z^0$ (no charge);

# The W and Z mesons and the weak interaction 161

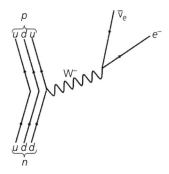

**Figure 13.8** The weak nuclear interaction, showing a neutron ($n$) in β-decay, in which a down quark ($d$) emits a $W^-$ particle, which in turn decays into an electron ($e^-$) plus antineutrino ($\bar{\nu}_e$); (p, proton; $u$, up quark.)

β-decay is mediated by the $W$ or the $Z$ bosons. These particles decay rapidly; so, for example, a neutron emits a $W^-$ to become a proton, and the $W^-$ decays quickly into an electron and antineutrino to produce the observed β-decay of the neutron. Figure 13.8 shows a down quark changing to an up quark by emitting a $W^-$. These bosons are massive particles (about 90 times the mass of the proton), as is consistent with the very short range of the interaction. The exchange of the neutral particle $Z^0$ between two particles leaves their identities unchanged.

Steven Weinberg, Abdus Salam, and Sheldon Glashow developed a theory that unifies the weak and the electromagnetic forces, giving it the title *electroweak interaction*. The mediating bosons are in a related grouping of the $W$ bosons, the $Z$ bosons, and the γ bosons (photons). But the photon is massless, so how can it be related to the others? At extremely high energies, perhaps soon after the Big Bang, all these bosons would have had zero mass; but, at lower energies, a new field, the Higgs interaction (discussed briefly in Sec. 13.8) gives the $W$ and $Z$ mass, but not the photon. The process is somewhat analogous to the phase transition in a ferromagnet. Above a certain temperature, the spins point randomly, but, below that transition temperature, the spins line up, choosing one particular direction out of all the possible directions in 3D space. Here, below a certain energy, the $W$s and the $Z$s become massive via the Higgs field.

We have mentioned that the electron and neutrino are particles involved in the weak interactions. These are two members of the class of particles known as *leptons*, that is, low-mass particles. Also in this group are the μ (muon) and the τ (tau). Each of these has an associated neutrino, $\nu$. The electron neutrino is involved in β-decay, but there is also the tau neutrino, $\nu_\tau$, as well as the muon neutrino, $\nu_\mu$, plus all their antiparticles. It is no wonder the array of all the elementary particles is sometimes called the "particle zoo."

## 13.8 The particle zoo

Figure 13.9 shows the presently known elementary particles according to the theory known as the Standard Model. Readers wanting more information on particle physics will find many popular books that go far beyond our brief outline here. The most recent addition to this list of particles is the Higgs boson. The Higgs field gives mass to all the elementary particles via its field. The Nobel Prize for the theory that predicted this behavior went to François Englert and Peter Higgs in 2013. A real Higgs boson can be generated at high energy, and this particle was discovered recently at the CERN particle accelerator known as the Large Hadron Collider (LHC). The Higgs particle is about 1.5 times as massive

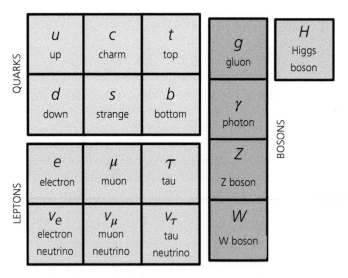

**Figure 13.9** The particles of the Standard Model.

as the $W$ particles and could be made only when the LHC was constructed to reach that energy. The Higgs causes mass by interacting with individual particles, but if it is exchanged, it creates a force. Because of the high mass of the particle, the force is very short range and is weaker than the electromagnet force. It may be difficult to detect directly.

Theories that extend the Standard Model include many more particle possibilities. In particular, in the supersymmetry (SUSY) model, each known fermion in nature would be paired with a boson, and each boson with a fermion. For example, the superpartner of the quark is named the "squark," and that of the electron is the "selectron." The photon partner is the "photino," and the gluon partner is the "gluino." None of these new particles have yet been observed.

## 13.9 Forces in condensed matter physics

We started this discussion by considering two bowling balls attracting each other by deforming the mattress on which they have been placed. This is very analogous to how some forces act in the theory of liquids and solids. When an electron moves in a certain kind of crystal known as a dielectric solid, it can pull the positive nuclear charges of the surrounding atoms closer to it and repel the negative electron charges to be a bit farther away, thus screening its own charge. The electron effectively gets an increased "effective mass" and moves more ponderously then. The *quasiparticle* thus formed is known as a *polaron*, since it has polarized the space surrounding it. This effect is analogous to the vacuum polarization we saw in Sec. 13.4, except here, the polarization is of real atoms.

Electrons in some metals will also pull the nearby positive charges closer, causing a positive charge cloud that can attract another electron; this is basically one electron interacting with another by exchanging a virtual phonon, a quantized sound wave, since a sound wave in a solid is a density distortion of the lattice. This attraction can cause the electron pair to bind together, forming what is called a *Cooper pair*. The cooperative pairing of many Cooper pairs results in a superconducting state in certain metals, as we have mentioned in Chap. 9.

If one dissolves $^3$He atoms in liquid $^4$He, two $^3$He atoms can exchange a virtual phonon in the liquid. Such an effective force has been observed in many experiments, making this system a beautiful example of an interacting dilute Fermi gas. We discussed this system briefly back in Chap. 11.

There are many analogies between particle physics and condensed matter physics. Even the Higgs mechanism of giving mass to the elementary particles was discovered first in the theory of superconductivity by Philip Anderson.

# 14

# Teleportation of a Quantum State

> If it was that easy, the American military would have figured it out years ago.
>
> CHRISTOPHER RANKIN

Teleportation is a science-fiction mode of travel used in the TV show *Star Trek*. A person stands in a "teleporter" on the starship *Enterprise*; he or she then disappears from the area and reappears elsewhere on a nearby planet or spaceship. Of course, the person can also be teleported back to the *Enterprise*. Quantum teleportation is not for people—at least, not yet. However, it is possible to "teleport" a quantum state from one place to another by using the properties of entanglement. Suppose Alice has a particle in a certain quantum state; she doesn't even have to know what the state is. Bob wants to have a particle in the identical state. Alice might just send an atom in that state to him by FedEx, for example, but perhaps this is not feasible or too slow. And, probably more importantly, even the gentlest FedEx handling could cause the superposed state to decohere. Teleportation is a way for her to transfer the state to Bob; he can end up with an atom the same state without even knowing what the state is. But it is guaranteed to be identical to what Alice had.

## 14.1 How it is done

The details of the teleportation process are somewhat mathematically arcane for a beginner in quantum mechanics, but it is worth going through it because the result is fascinating, and the methods used are typical of standard quantum techniques. In teleportation we need a "change of basis" to prepare a wave function for a particular experimental arrangement. We have done this before, in Chaps. 5 and 12, but let's see in detail what we mean in the simpler situation of a single spin. We suppose we have a spin state in a superposition of spin states along $x$:

$$\Psi = c\psi_{\uparrow x} + d\psi_{\downarrow x}. \tag{14.1}$$

This state has a probability of $c^2$ of having an up spin along $x$, and a probability of $d^2$ of the spin being down along $x$ when we measure it in an SGA that is aligned along $x$. But suppose we intend to measure spin along $z$. Then we should express the wave function in a proper "set of basis states" for this different measurement. We use the expressions from Eqs. (5.10) and (5.11) of Sec. 5.3, as these tell us how to make the changes:

$$\psi_{\uparrow x} = \frac{1}{\sqrt{2}} \left( \psi_{\uparrow z} + \psi_{\downarrow z} \right), \text{ and} \qquad (14.2)$$

$$\psi_{\downarrow x} = \frac{1}{\sqrt{2}} \left( \psi_{\uparrow z} - \psi_{\downarrow z} \right). \qquad (14.3)$$

Substitute these for the two $x$ wave functions in Eq. (14.1) to get

$$\Psi = \left( \frac{c+d}{\sqrt{2}} \right) \psi_{\uparrow z} + \left( \frac{c-d}{\sqrt{2}} \right) \psi_{\downarrow z}. \qquad (14.4)$$

The probability of finding the spin up along $z$ is $\left( \frac{c+d}{\sqrt{2}} \right)^2$, and that of finding the spin down along $z$ is $\left( \frac{c-d}{\sqrt{2}} \right)^2$. We have done a "change of basis states" to account for the experiment. The wave function is the same one we started with, but we did a mathematical change to account for the planned experimental set up. A wave function is very versatile and can accommodate description of a wide variety of experiments, but as we have stressed before, the experiment determines the most convenient basis in which to work, in this sample case, the one with spins along $z$.

Now let's apply the general idea of a change of basis to teleportation. We suppose Alice has a spin state for particle number 1 given by

$$\phi(1) = a\psi_{\uparrow}(1) + b\psi_{\downarrow}(1), \qquad (14.5)$$

(where the arrows now represent up and down along $z$, without writing "$z$"). So, this spin has probability $a^2$ of being up along the $z$-axis, and $b^2$ of being down along the axis. This is the state Alice wishes to teleport to Bob. (The particle does not have to be an electron—any particle with

# How it is done

spin will do, or photons with their two polarization states—just write H for ↑, and V for ↓—or certain atoms that have just two states.) We simplify the notation because we have to write it a lot:

$$\phi(1) = a|+_1\rangle + b|-_1\rangle. \tag{14.6}$$

The notation $|+_1\rangle$ means particle 1 with spin up; $|-_1\rangle$ means particle 1 with spin down. (The strange notation of $|X\rangle$ for a wave function was invented by Paul Dirac for a mathematical reason that we will not explain here. We actually already used the notation a bit in Chap. 13.)

Alice also has a source of spins that sends out two other particles, numbers 2 and 3, one with spin up and one with spin down, with 2 going to Alice, and 3 sent off to Bob, as in Fig. 14.1. The spins are in a singlet state (Eq. [5.8]) as in our Bell theorem analysis. That is, we have

$$\psi(2,3) = \frac{1}{\sqrt{2}} \left[ \psi_\uparrow(2)\psi_\downarrow(3) - \psi_\downarrow(2)\psi_\uparrow(3) \right]$$

$$= \frac{1}{\sqrt{2}} (|+_2-_3\rangle - |-_2+_3\rangle), \tag{14.7}$$

with the two particles clumped together in the Dirac state, that is, $|+_2-_3\rangle$ means the same as $\psi_\uparrow(2)\psi_\downarrow(3)$. (Note that when two particles do not interact, we can write their joint wave functions as a product like $\psi_\uparrow(2)\psi_\downarrow(3)$. Even though the particles do not interact, they can still

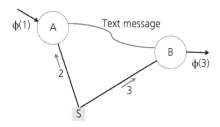

**Figure 14.1** The experimental arrangement for teleporting a particle's state from Alice to Bob. At A, Alice has particle 1 in the state $\phi(1)$ to be teleported. Particles 2 and 3 are produced at S in a singlet state; particle 2 goes to Alice, and particle 3 to goes Bob (as indicated by the arrows). Alice does a "Bell-state" experiment on particles 1 and 2. She sends the result to Bob via a text message, which tells him which rotation operation he should apply to particle 3 to put it into the same state that particle 1 originally had.

prepared in a superposition of states like $\psi(2, 3)$, where we do not know which of the two states ($|+_2-_3\rangle$ or $|-_2+_3\rangle$) the particles have, because they have both.) Now, there are three particles altogether, and the joint wave function for all three is

$$\Psi_{\text{All}}(123) = \phi(1)\psi(23) \qquad (14.8)$$
$$= \frac{a}{\sqrt{2}} (|+_1 +_2 -_3\rangle - |+_1 -_2 +_3\rangle)$$
$$+ \frac{b}{\sqrt{2}} (|-_1 +_2 -_3\rangle - |-_1 -_2 +_3\rangle),$$

where we have simply multiplied out the product of the qualities in Eqs. (14.6) and (14.7) and combined their Dirac states (see Fig. 14.1 for the situation so far).

Alice has access to particles 1 and 2 while Bob is receiving particle 3. Alice is going to do an experiment on her two particles that will collapse Bob's particle's state into one almost identical to the state $\phi(1)$ that Alice had originally. To do this, Alice is going to measure in which of the following rather complicated states her particles 1 and 2 appear. That is, she selects a set of new basis states for her measurements on the two particles 1 and 2; these are called "Bell states":

$$\Psi_-(12) = \frac{1}{\sqrt{2}} (|+_1-_2\rangle - |-_1+_2\rangle), \qquad (14.9)$$

$$\Psi_+(12) = \frac{1}{\sqrt{2}} (|+_1-_2\rangle + |-_1+_2\rangle), \qquad (14.10)$$

$$\Phi_-(12) = \frac{1}{\sqrt{2}} (|+_1+_2\rangle - |-_1-_2\rangle), \text{ and} \qquad (14.11)$$

$$\Phi_+(12) = \frac{1}{\sqrt{2}} (|+_1+_2\rangle + |-_1-_2\rangle). \qquad (14.12)$$

These states form a "complete set," which means *any* two-particle spin state can be expressed in terms of them. Note that the first of these, Eq. (14.9), is just a singlet state. Just how Alice can measure which of these states she finds her pair of particles to have in a experiment is not immediately obvious, and we will discuss it in Sec. 14.2.

But to prepare for this experiment we must rewrite our three-particle wave function of Eq. (14.8) in terms of these states—this is just analogous to the change from Eq. (14.1) to Eq. (14.4). The result is

$$\Psi_{\text{All}}(123) = \frac{1}{2}\big[-\Psi_-(12)(a\,|+_3\rangle + b\,|-_3\rangle) \qquad (14.13)$$
$$-\Psi_+(12)(a\,|+_3\rangle - b\,|-_3\rangle)$$
$$+\Phi_-(12)(a\,|-_3\rangle + b\,|+_3\rangle)$$
$$+\Phi_+(12)(a\,|-_3\rangle - b\,|+_3\rangle)\big].$$

You should now substitute Eqs. (14.9)–(14.12) into this equation, multiply it all out, and verify that there are the necessary cancellations and additions for this equation to become identical with Eq. (14.8); doing that should ensure that you understand what has been done here. (Note, e.g., that $|+_1-_2\rangle\,|+_3\rangle$ is the same as $|+_1-_2+_3\rangle$.)

When Alice does her measurement on particles 1 and 2, she might find them to be in state $\Psi_-(12)$. Then, the three-particle wave function has collapsed to the first term of Eq. (14.13), and if Alice tells Bob what she got, then he knows immediately that his particle-3 state is $(a\,|+_3\rangle + b\,|-_3\rangle)$, which is exactly the particle-1 state Alice started with! (The minus sign out front of the $\Psi_-(12)$ term in Eq. [14.13] is of no importance, since probabilities involve squaring the wave function.) The original state of particle 1 has been teleported to Bob by putting his particle 3 in the same state that particle 1 was in. But what if Alice's measurement yields one of the three other possible Bell states?

Alice can use email, a telephone, or some other normal classical communication system to tell Bob what Bell state she got for particles 1 and 2: $\Psi_-$, $\Psi_+$, $\Phi_-$, or $\Phi_+$. In Fig. 14.1, she uses a text message. Suppose, for a simple case, we had $a = b = 1/\sqrt{2}$, in which case the original spin would have been up along $x$ (see Eq. (14.2)), and Alice finds her experiment gives state $\Psi_+(1,2)$. Then, by Eq. (14.13), Bob's particle 3 has collapsed into state $\frac{1}{\sqrt{2}}(|+_3\rangle - |-_3\rangle)$, which is a spin pointing along $-x$ (by Eq. [14.3]). So, to get the original state, Bob applies a magnetic field for a time such that the spin rotates $180°$ to up along $x$; and he now has the original state. In this case, he did a rotation around the $z$-axis. In each case, whatever Alice gets, he will know exactly what definite rotation to apply to transform the state he got into the teleported state. If she gets $\Psi_+$, he always rotates $180°$ around the $z$-axis—no matter what $a$ and $b$ are; in the case of $\Phi_-$, he always rotates $180°$ around $x$; and for $\Phi_+$, around $y$.

Alice no longer has the original state. It has been teleported away to Bob. This satisfies another theorem of quantum mechanics: one cannot, without knowing what the state is, clone a state, that is, simply make a duplicate of it; if you try, the original will be destroyed. Note the four steps in teleportation: (1) Alice keeps particle 2 while sending particle 3, which is entangled in a singlet state with particle 2, to Bob; (2) Alice makes a measurement on the pair state of particles 1 and 2; (3) Alice sends a classical communication of her measurement result to Bob; and (4) Bob does a rotation based on Alice's result to produce the correct state.

Anton Zeilinger, one of the people who invented teleportation, has been involved in several experiments in which a photon, in a particular polarization state, is teleported over large distances. I recommend his popular book *Dance of the Photons*. Recently, Chinese physicists performed a teleportation of the state of a collection of rubidium atoms to another collection 150 meters away. The world record long-distance teleportation is by Ma et al. from the Institute for Quantum Optics and Quantum Information in Vienna Austria; they teleported a state 143 km—from one of the Canary Islands to another.

## 14.2 Measuring Bell states

Alice needs to make a measurement to determine in which of the four Bell states her 1–2 pair is found. Any pair wave function can be expressed as a sum over these four states, and, in the Bell measurement, it will collapse to one of them. We can use results from Sec. 11.4 to show how to distinguish at least one such state. We note from Eq. (14.9) that $\Psi_-$ is antisymmetric; that is, if we interchange spins 1 and 2, one term becomes the other, and the wave function changes sign. The other three states are symmetric; if we interchange the spins there, we get the same thing back. Suppose our particles are bosons. They can be photons, for example, with the "spins" being polarization directions, as we indicated in Sec. 14.1. If the particles are bosons, the total wave function must be symmetric, so how can we have an antisymmetric pair state like $\Psi_-$? The answer is that the wave function shown is just the spin part; we need a space part also to describe the entire wave function: $\Psi_-$ just describes the probability of finding *spins* up or down and does not say what the probability is of finding the particles at certain *positions*. We considered such a situation back in Eq. (11.7) for fermions; in that case, we had an antisymmetric spin state multiplying a symmetric space state. So, what we need here is an antisymmetric space state part multiplying $\Psi_-$ to give an overall symmetric wave function:

$$\Psi_{\text{total}} = \Psi_-(1,2)\frac{1}{\sqrt{2}}\left[u(x_1)v(x_2) - v(x_1)u(x_2)\right], \quad (14.14)$$

where $u(x)$ and $v(x)$ are one-particle wave functions that describe the particle position $x$. If I interchange the pair of particles, $\Psi_-$ changes sign because the spin coordinates are interchanged, but the new space part also changes sign, as $x_1$ and $x_2$ are interchanged. The total wave function does not change sign, since the two minus signs cancel out. The three other Bell states have symmetric spin states: they must each have symmetric space states of the form $\frac{1}{\sqrt{2}}\left[u(x_1)v(x_2) + v(x_1)u(x_2)\right]$.

We know from Sec. 11.4 that two particles in symmetric space states sent through a beam splitter will end up on the *same* side of the beam splitter. In the case of an antisymmetric space state, particles end up on *opposite* sides. So if we find both particles 1 and 2 on the opposite side, then they must have been in Bell state $\Psi_-$. When Alice finds this result, she can tell Bob that his state is correct and needs no adjustment. Without further measurements, she could not distinguish among the other three. However, instead of having to set up some more elaborate experiment, Alice and Bob could just repeat the process; once out of every four tries, on average, they should get the antisymmetric space result implying the $\Psi_-$ state. And then Bob would have his desired state.

Actually, Alice can do a bit better if she can measure spin (or polarization). If both particles come out on the same side *and* she can measure the spins of the two particles, she can distinguish one other state. Which one is it?[1] Much more elaborate experimental arrangements allow one to distinguish all four Bell states.

Teleportation is potentially useful in quantum computing, as the transfer of quantum information is at the heart of quantum computing. A wave function in a superposition is a fundamental object in quantum computing, called a qubit. Exchanging information is necessary in computing and so teleporting qubits could be an important aspect.

---

[1] If Alice can measure spin, then when she gets two spins on the same side that are different, she knows that she has $\Psi_+(12)$, which symmetrically entangles opposite spins, as seen in Eq. (14.10). But if they are the same, the result is either $\Phi_+$ or $\Phi_-$, but she cannot tell which.

# 15

# Quantum Computing

> Even a child of five could understand this. Someone go fetch a child of five.
>
> GROUCHO MARX

One of the more speculative and popular applications of quantum mechanics is quantum computing. Government agencies and corporations, as well as scientists, are engaged in active research in this area. A quantum computer has the potential to break very sophisticated security codes, if it can be successfully developed. But, given the inherent difficulties in keeping quantum systems from decohering, it may be a while before more than just the simplest quantum computers are built.

A classical computer is based on manipulating bits, 0s or 1s, which are represented by states of computer memory, and, for example, might be magnetic material polarized either up or down. Such bits can represent any number by using the binary number system. Recall what the decimal system means. In a decimal number, a digit in the $n$th column represents 0 through 9 times $10^n$; one then adds the results of all the columns to make the complete number. Thus, we have $362 = 3 \times 10^2 + 6 \times 10^1 + 2 \times 10^0$, where $10^0 = 1$. The binary system uses only the digits 0 and 1. A digit in the $n$th column represents 0 or 1 times $2^n$; adding the results of all the columns make the complete number. Thus, $101 = 1 \times 2^2 + 0 \times 2^1 + 1 \times 2^0$ is 5, when translated into the decimal system. Table 15.1 shows the representation of the first eight digits, 0 to 7.

A memory unit, a bit, in a classical computer can be either 0 or 1. But a quantum state of an atom can be in a superpositions of states:

$$|\psi\rangle = \alpha |0\rangle + \beta |1\rangle, \qquad (15.1)$$

where we have to have $\alpha^2 + \beta^2 = 1$ because the coefficients are related to probabilities. We have again used the Dirac notation for our bit

Quantum Computing

**Table 15.1** Digits 0 to 7, in binary form

| Digit | Binary form |
|---|---|
| 0 | 000 |
| 1 | 001 |
| 2 | 010 |
| 3 | 011 |
| 4 | 100 |
| 5 | 101 |
| 6 | 110 |
| 7 | 111 |

states: $|0\rangle$ or $|1\rangle$. Our memory bits can be made up of atoms that each have two energy levels; an atom in the lower energy level corresponds to $|0\rangle$, and an atom in upper energy level corresponds to $|1\rangle$. The atom is in both a 0 state and a 1 state, simultaneously. The atomic state acting as a quantum memory device in such a state is called a *qubit*. Can we exploit this feature to do faster or more complicated calculations? Yes, we can, but it is not so easy.

The math notation in this chapter, while no worse than a form of algebra, can look pretty intense. However it is not that complicated until we get to Sec. 15.4, where it does get more serious.

A computer has an input register and an output register such that an operation on the bits in the input produces the output. An operation **X**, called a NOT operator, flips the bit:

$$\mathbf{X}|0\rangle = |1\rangle, \text{ and} \quad (15.2)$$
$$\mathbf{X}|1\rangle = |0\rangle.$$

Using several atoms, we can compose a more complicated number made up of several such Dirac bits, say $|1\rangle_1 |0\rangle_2 |1\rangle_3 = |101\rangle$, where the last form is shorthand for the first. An operator that acts on the whole can be made up of individual operators:

$$\mathbf{X}_1 \mathbf{I}_2 \mathbf{I}_3 |010\rangle = |110\rangle, \quad (15.3)$$

where **I** is the unit operator that does not change the bit state. In general, a computer operation acts on a number made up of bits $|xyz\rangle$ and produces some other number $|f(xyz)\rangle$, where $f$ is whatever the operator makes out of $xyz$:

$$\mathbf{F}|xyz\rangle = |f(xyz)\rangle. \tag{15.4}$$

A classical computer does one operation at a time on a fixed input register to produce the unique output for the operation.

The state $|110\rangle$ has two atoms in their excited states, and one in the ground state, and could represent the number 6 in binary. To do an operation, we need to be able to change the states of the atoms; this is possible in a variety of ways. Consider the effect of the superposition shown in Eq. (15.1). Having three atoms each in a superposition with $\alpha = \beta = 1/\sqrt{2}$ would correspond to the entangled state

$$\begin{aligned}|\psi\rangle &= \left(\frac{1}{\sqrt{2}}\right)^3 (|0\rangle_1 + |1\rangle_1)(|0\rangle_2 + |1\rangle_2)(|0\rangle_3 + |1\rangle_3) \\ &= \left(\frac{1}{\sqrt{2}}\right)^3 (|000\rangle + |001\rangle + |010\rangle + |011\rangle \\ &\quad + |100\rangle + |101\rangle + |110\rangle + |111\rangle).\end{aligned} \tag{15.5}$$

That is, the quantum state of the $n$ atoms can represent *all* the numbers from 0 to $2^n - 1$ simultaneously! An operation **F** on this state gets all the possible results of the operation on *any* $n$-digit number simultaneously. This remarkable ability is known as *quantum parallelism*.

Maintaining quantum coherence, that is, the preservation of the superposition states of all the atoms and their interference capabilities, while this computation is going on is absolutely necessary. Otherwise, we lose the parallel computations. But it is precisely interaction with the external world that we need to read out the result of any computation. When we do measure the quantum output register after the operation **F** is finished, we collapse the wave function to just one of the results! We have learned no more than we would have with a classical computer, and perhaps less, since we may not know which input bit corresponds to the output bit we found.

A further difficulty in quantum computing is error correction. One technique used in a normal computer is repetition, in which a string of data might be sent three times. If one string is seen to be different, then we know an error has been made. However, quantum states cannot be duplicated. (In teleportation, a state is transferred to a different place, but the original state is lost.) Moreover, any kind of state checking by measurement collapses the state, losing the quantum parallelism. Fortunately, quantum error correction advances are still being made.

Despite these problems, there is still hope for quantum computing to be useful. While we cannot find out each value of each $|f(xyz)\rangle$, we might find the *relationship* between, say, $|f(000)\rangle$ and $|f(110)\rangle$. And, indeed, it turns out to be possible to find such relationships in a computational time that no classical computer could match. Let's look at some example problems.

## 15.1 Deutch's problem

The solution to this problem likely has no particular use but was invented by John Deutch basically to show that you could do *something* with a quantum computer faster than on a classical computer. You are given a black box that performs one of the four simple computations shown in Table 15.2.

$$\mathbf{F}_i |x\rangle = |f_i(x)\rangle \tag{15.6}$$

Thus, for example, $\mathbf{F}_0 |x\rangle = |0\rangle$ for both $x = 0$ or 1, while $\mathbf{F}_2 = \mathbf{X}$ (NOT) always flips the bit ($\mathbf{F}_2 |0\rangle = |1\rangle$; $\mathbf{F}_2 |1\rangle = |0\rangle$). We do not know which of the four operators is in our black box. We can allow it to operate just once. All we ask is to learn whether the operator produces *identical output*

Table 15.2  Output from the black box

| Operator | Output | |
|---|---|---|
|  | $\mathbf{F}_i |0\rangle$ | $\mathbf{F}_i |1\rangle$ |
| $\mathbf{F}_0$ | $|0\rangle$ | $|0\rangle$ |
| $\mathbf{F}_1$ | $|0\rangle$ | $|1\rangle$ |
| $\mathbf{F}_2$ | $|1\rangle$ | $|0\rangle$ |
| $\mathbf{F}_3$ | $|1\rangle$ | $|1\rangle$ |

bits, as do $F_0$ and $F_3$, or *different output bits*, like $F_1$ and $F_2$. Can we do this with a classical computer in one operation? Suppose I let the operator act on a $|0\rangle$ and I get $|1\rangle$; then I know only that the operator is either $F_2$ or $F_3$, one of which gives the same results on both inputs, and the other, different. I need to let the operator act a second time with the classical computer.

But with a quantum computation, I can do it on one step. You can either take my word for it or work through the set of operations outlined in Sec. 15.4. It is not as quite as complicated as it looks there at first glance!

## 15.2 Grover's search algorithm

Lov Grover found a way to use a quantum computer to search a set of data to find the one item that satisfies some specified rule. For example, suppose we want to find the name associated with a given phone number $x$ in a normal phone book listed by name alphabetically. If there are $N$ entries, we would have to search $N/2$ of them to have a 50 % probability of finding both $x$ and the name associated with $x$. Grover's method allows the quantum computer to find the correct number, with almost certain probability, in about $\sqrt{N}$ steps, which is much a much smaller number of steps than $N/2$ if $N$ is large.

## 15.3 Shor's period-finding algorithm

One of the most important encryption schemes, the RSA method, named after its inventors Ron Rivest, Adi Shamir, and Leonard Adleman, who first publicly described the algorithm in 1977, uses the product of two very large prime numbers, $p$ and $q$, to devise a code key. Breaking the code would require the equivalent of finding the factors $p$ and $q$ of this large number. Finding the factors in products of large prime numbers turns out to depend on being able to find the period of a periodic function. A function $f(x)$ is periodic with period $r$ if $f(x + nr) = f(x)$, where $n$ is any integer; each time you go a distance $r$, you find $f$ has the same value. The best period-finding method programmed on a classical computer to factor a number having $n$ binary bits requires something like $e^{n^{1/3}}$ operations—that is, basically exponentially increasing with length. However, Peter Shor discovered a method that, on a quantum computer, would require only on the order of $n^3$ operations. The development of quantum computers is therefore likely of great interest to organizations like the National Security Agency and banks, which do not want their security codes broken! The method has been tested on very small quantum computers, which were able to prove that $15 = 5 \times 3$! There are other problems that have been found that can be solved by quantum computer, but the Shor solution is probably the most important by far.

## 15.4 The solution to Deutch's problem

If you want to understand in more detail how a quantum computer might solve a problem, I give here the solution to Deutch's problem.

### 15.4.1 Preliminaries

In Eq. (15.2), we showed a NOT operation on the input register. We need another operation, the "Hadamard operator," **H**, which puts the input register into a superposition of states:

$$\mathbf{H} |0\rangle = \frac{1}{\sqrt{2}} (|0\rangle + |1\rangle), \text{ and} \qquad (15.7)$$

$$\mathbf{H} |1\rangle = \frac{1}{\sqrt{2}} (|0\rangle - |1\rangle). \qquad (15.8)$$

Operations like this are not hard to implement on atom or spin states, for example. Use these definitions to prove for yourself that

$$\mathbf{H} \frac{1}{\sqrt{2}} (|0\rangle + |1\rangle) = |0\rangle, \text{ and} \qquad (15.9)$$

$$\mathbf{H} \frac{1}{\sqrt{2}} (|0\rangle - |1\rangle) = |1\rangle. \qquad (15.10)$$

In computing, the contents of the input register are not just replaced by those of the output register; both are kept. So, we should include *both* registers on both sides of the equation in the form $|input\rangle |output\rangle$. If the operation is **F** and we assume the input register is initially $|x\rangle$ and the output initially set at $|0\rangle$, then the operation looks like

$$\mathbf{F} |x\rangle |0\rangle = |x\rangle |f(x)\rangle. \qquad (15.11)$$

Thus, **F** acts on the input, $|x\rangle$, and changes the initial output $|0\rangle$ into $|f(x)\rangle$. If the output is already $|1\rangle$, then we need to add $f(x)$ to it:

$$\mathbf{F}|x\rangle|1\rangle = |x\rangle|1 \oplus f(x)\rangle, \tag{15.12}$$

where the addition is done modulo 2, that is, while $|1 \oplus 0\rangle = |1\rangle$, going over 1 by 1 reverts to 0: $|1 \oplus 1\rangle = |0\rangle$.

There is no reason why we might not prepare the input or output independently of one another before applying the operator **F**. We might even prepare both the input and the output simultaneously by applying 1-qubit operators to each. For example,

$$\mathbf{H} \times \mathbf{X}|0\rangle|0\rangle = \left[\frac{1}{\sqrt{2}}(|0\rangle + |1\rangle)\right]|1\rangle = \frac{1}{\sqrt{2}}(|0\rangle|1\rangle + |1\rangle|1\rangle), \tag{15.13}$$

where we separate the two operators for the input and output states by a times sign. The input and output registers now represent an *entangled state*. Let's now apply the $\mathbf{F}_3$ operator from Table 15.2. This operation adds a 1 in the output register, no matter what was in the input register. Thus,

$$\mathbf{F}_3 \frac{1}{\sqrt{2}}(|0\rangle|1\rangle + |1\rangle|1\rangle) = \frac{1}{\sqrt{2}}(|0\rangle|0\rangle + |1\rangle|0\rangle). \tag{15.14}$$

Note that, in this case, where the output already registers a $|1\rangle$ state, we had to add in the 1, modulo 2: $\mathbf{F}_3|0\rangle|1\rangle = |0\rangle|1 \oplus 1\rangle = |0\rangle|0\rangle$. In order to introduce some necessary general notation, we write the operation in the following way:

$$\mathbf{F}_3 \frac{1}{\sqrt{2}}(|0\rangle|1\rangle + |1\rangle|1\rangle) = \frac{1}{\sqrt{2}}\left(|0\rangle\left|\widetilde{f_3}(0)\right\rangle + |1\rangle\left|\widetilde{f_3}(1)\right\rangle\right). \tag{15.15}$$

In the first term, the input is 0, so the output is $f_3(0) = 1$; but there is already a 1 in the output register so we get $f_3(0) \oplus 1$, which we call $\widetilde{f_3}(0)$. In the second term, we get $f_3(1) \oplus 1$, which we call $\widetilde{f_3}(1)$. In the case of $\mathbf{F}_3$, we have $\widetilde{f_3}(0) = \widetilde{f_3}(1) = 1 \oplus 1 = 0$, as we wrote in Eq. (15.14).

### 15.4.2 The solution

Now, we are ready to attack the problem. Prepare the input/output in $|1\rangle|1\rangle$, and then apply a double Hadamard:

## The solution to Deutch's problem

$$\mathbf{H} \times \mathbf{H} |1\rangle |1\rangle = \left[\frac{1}{\sqrt{2}}(|0\rangle - |1\rangle)\right]\left[\frac{1}{\sqrt{2}}(|0\rangle - |1\rangle)\right] \quad (15.16)$$

$$= \frac{1}{2}(|0\rangle |0\rangle - |1\rangle |0\rangle - |0\rangle |1\rangle + |1\rangle |1\rangle).$$

Apply the unknown $\mathbf{F}_i$ operator to this, and use the $\widetilde{f}(x)$ notation. We apply $\mathbf{F}_i$ to each term:

$$\mathbf{F}_i(\mathbf{H} \times \mathbf{H} |1\rangle |1\rangle) \quad (15.17)$$

$$= \frac{1}{2}\left(|0\rangle |f(0)\rangle - |1\rangle |f(1)\rangle - |0\rangle |\widetilde{f}(0)\rangle + |1\rangle |\widetilde{f}(1)\rangle\right).$$

We had to use the $\widetilde{f}$ notation in the last two terms because the output state was already a $|1\rangle$.

What we want to know is whether $f(0) = f(1)$. If, in Eq. (15.17), we have $f(0) = f(1)$, then we also have $\widetilde{f}(0) = \widetilde{f}(1)$, and this output state factors to become

$$\frac{1}{2}\left[(|0\rangle - |1\rangle)\left(|f(0)\rangle - |\widetilde{f}(0)\rangle\right)\right], \quad \text{for } f(0) = f(1). \quad (15.18)$$

On the other hand, if $f(0) \neq f(1)$, then $\widetilde{f}(0) = 1 \oplus f(0) = f(1)$, and $\widetilde{f}(1) = 1 \oplus f(1) = f(0)$, since 1 and 0 are the only two choices for the two $f$s. Then, the result is

$$\frac{1}{2}\left[(|0\rangle + |1\rangle)\left(|f(0)\rangle - |\widetilde{f}(0)\rangle\right)\right], \quad \text{for } f(0) \neq f(1). \quad (15.19)$$

The final step is to apply a Hadamard transformation to the input register. By Eqs. (15.9) and (15.10), we get the following for the two cases:

$$|1\rangle \frac{1}{\sqrt{2}}\left(|f(0)\rangle - |\widetilde{f}(0)\rangle\right), \quad \text{for } f(0) = f(1), \text{ and} \quad (15.20)$$

$$|0\rangle \frac{1}{\sqrt{2}}\left(|f(0)\rangle - |\widetilde{f}(0)\rangle\right), \quad \text{for } f(0) \neq f(1). \quad (15.21)$$

We now simply measure the input register to determine the answer to the problem. If it is $|1\rangle$, then the two $f$s are equal; if $|0\rangle$, they are unequal. We had to use the operation of the unknown function in the computation only once. When one measures the output register, the state of the output register collapses to either $|f(0)\rangle$ or $|\widetilde{f}(0)\rangle$, that is, $|0\rangle$ or $|1\rangle$, with equal probability, giving us no further information.

## 15.5 Is anything built yet?

Only small quantum computers have been built to date. Maintaining entangled states coherently for long times has limited the size possible. A company called D-Wave Systems has sold what is claimed to be a 1000-qubit quantum computer. It uses "quantum annealing" to find the global minimum of functions, as required in optimization problems. So, basically, it would seem to use quantum tunneling to visit many configurations of an energy surface to find the lowest energy. Whether it is a true quantum computer can be debated, since it is not clear that it ever exhibits any entanglement of computer memory elements, which, in this case, are coupled superconducting flux qubits.

A very good introduction to quantum computing is the book *Quantum Computer Science, An Introduction*, by N. David Mermin.

# 16

# Weird Measurements

*Measure what is measurable, and make measurable what is not so.*
GALILEO GALILEI

## 16.1 Weak measurement

Up to now, we have considered measurements that collapsed the wave function into a definite state of the quantity being measured. Thus, in the Stern–Gerlach experiment, the spin comes in as a superposition of up and down along some axis, and the magnet provides different momenta to each spin component, separating them into two distinct beams. A detector collapses the wave function into either the upper or the lower component beam. Such a detection procedure that picks out one of the possibilities is called a "projective measurement." Another example is the two-slit experiment. If we attempt to determine through which slit the particle passes with detectors at the slits, the superposition collapses to a single term and only one detector triggers. If we do not have detectors at the slits, then a single blip occurs on the screen, meaning the wave function has collapsed to that single position.

However, it is possible to have our system of interest interact so weakly with another measurement system that the superposition is not upset. For example, suppose our Stern–Gerlach magnet, set to determine spin in the $z$-direction, is very weak, and the beam of particles passing through the magnet has a finite width, that is, particles do not all pass along the exact same line but have a variety of paths within a circular cross section radius $r$. The weak magnet makes an individual path deviate by a distance $a$ that is less than $r$, so that we cannot actually distinguish the up and down spin paths. However, if a normally strong Stern–Gerlach measurement in the $x$-direction follows the weak $z$-measurement, and one examines the up-$x$ beam, one can find, as shown in 1988 by Aharonov, Albert, and Vaidman, that the pattern on the screen may be deviated in the $z$-direction by an amount proportional to many times $a$. The subsequent projective measurement on the weakly measured particle–wave function can result in an *amplification* of the result for the original weak measurement. The second strong measurement on the particle is only possible because the weak measurement has not collapsed the original wave function.

The mathematical details of this weak/strong measurement sequence are a bit too complicated to outline here, but we can look at how it is used in some rather unusual experiments. We first consider the two-slit experiment in which the one detects the average particle trajectories taken by particles, but without ruining the two-slit interference! The second experiment involves the actual measurement of a wave function. Each of these experiments is unusual because it is looking at something that had not been thought to be available to experiment. I think these qualify as weird measurements.

## 16.2 Measuring two-slit trajectories

Bohm's theory, which we discussed in Chap. 5, interprets quantum mechanics by positing actual particle paths governed by equations that invoke the wave function as an actual field that guides the particles in their paths. The particles have random initial conditions and so there are many different paths followed. On the other hand, in an alternative mathematical formulation of standard quantum mechanics, Richard Feynman has constructed what is known as the "path-integral" method. In this, the behavior of the wave function is derived by averaging over all classical paths between a position at one time, and that at a later time. Thus, the experiment we describe here does not prove, by showing particle trajectories, that an alternative interpretation of quantum mechanics is favored. It is not following a single particle along a definite path but showing some average momentum direction at each transit between slits and screen.

The experimental arrangement used by coworkers from universities in Canada, Australia, and France is shown schematically in Fig. 16.1. The idea is to determine the momentum vector of the particles at each point in their flight from near the slits to the screen. From the momentum, one can determine the average direction of motion (i.e., trajectory) of the photons.

Photons pass through the slits and through a polarizer at a 45° angle, which leaves them in a superposition of equal horizontal and vertical

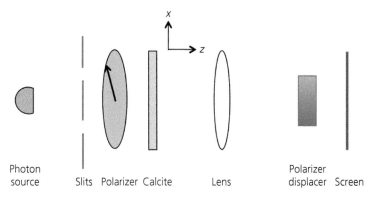

**Figure 16.1** Measuring the trajectories of particles in the two-slit experiment. Photons pass through the slits, are polarized at 45°, and then have their polarization slightly shifted by an angle dependent on their x-momentum by a birefringent calcite crystal (see text). The moveable lens then focuses a particular z-plane onto a screen that detects the interference pattern as it is at that value of z. Before reaching the screen, the polarization is spatially separated into two components so that two interference patterns appear separated by a small y displacement. The difference in the intensity of the two patterns determines the polarization angle shift.

polarizations:

$$\Psi = \frac{1}{\sqrt{2}} \left( |H\rangle + |V\rangle \right). \tag{16.1}$$

The next element is the calcite crystal, which is *birefringent*. We need a slight diversion here to discuss birefringent crystals. When light enters any material, its propagation is slowed as it is passed along from atom to atom through the substance. Any light ray entering the crystal at an angle is *refracted*, that is, slightly changed in direction, by this change in velocity, as shown in Fig. 16.2. However, with a calcite crystal, the propagation velocity of the ray depends on polarization. The polarization parallel to the so-called optic axis of the crystal (see Fig. 16.2) is slowed more than that of the perpendicular component and so suffers more refraction. The amount of refraction depends on the angle of entry of the light, that is, on the x-component (i.e., the component parallel to the surface of the crystal) of the momentum of the photon. Because of the different ray velocities and the different path lengths in the crystal, the relative phase between the two rays is shifted (the places where wave maxima occur no longer coincide) so the polarization of

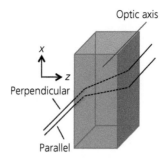

**Figure 16.2** A light ray passing through a birefringent calcite crystal. Light entering at the left is refracted (deviated in angle) by the crystal. Different polarizations of light travel at different velocities in the crystal, with the polarization parallel to the crystal optic axis refracted more than the polarization perpendicular to the optic axis. The result is a shift in the polarization angle of the outgoing ray. The input ray deviations from horizontal (parallel to $z$) are greatly exaggerated here.

the outgoing ray is shifted by a very small angle $\phi$ that is proportional to $p_x$, the $x$-component of the photon's momentum. (The outgoing photon is now *circularly* polarized, but we are not going into detail here on exactly what this means.) Through this weak measurement, we have encoded the $p_x$ momentum into a photon polarization angle. *The deviation caused by the calcite crystal is so small that the interference pattern is not destroyed.* This is a weak measurement. The strong final measurement is that of the position of the photon on the screen.

The beam is focused by the lens system and then passed through what we call a "polarization displacer." This element separates the beam into two polarization components, which are displaced slightly from one another in the $y$-direction (perpendicular to the plane of the figure); the polarization rotation angle is determined by the intensity difference between the two components. Each polarization component shows a diffraction pattern. The lens imaging system (actually three lenses) selects which $z$-plane is to be examined. This is equivalent to moving the screen to see the form of the diffraction pattern at various $z$-distances from the slits.

Because the average $x$-component and the total momentum of a photon are then known from the polarization rotation angle, one has the momenta in both relevant directions, $x$ and $z$. (The momentum is measured at the calcite crystal, but momentum is conserved and so is

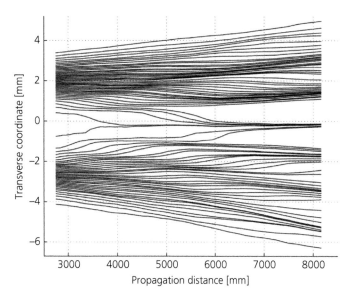

**Figure 16.3** Average photon trajectories in a two-slit interference experiment. A sample of 80 trajectories is shown. The "propagation distance" is what we called $z$ in the discussion. The interference pattern is evident in the trajectories. (From S. Kocsis et al, Science 332, 1170 (2011). Reprinted with permission from AAAS.)

the same where the position is measured, i.e., at the lens.) The momenta give the direction of travel at each $x$ and $z$, thus establishing the trajectory. The diffraction patterns are allowed to accumulate many photons (about 31,000) in data at 41 different $z$ planes. The experiment is not about seeing the trajectory of any single photon, but about seeing an average of very many photons at each position. The results are shown in Fig. 16.3. The trajectories clearly show the paths from the slit, and the formation of the interference fringes.

## 16.3 Measuring a wave function

Weak measurements can also explicitly measure a wave function. Such a result is distinct from simply measuring the probability of finding a particle at a position as given by the wave function squared. The wave function describes a wave, so it has both an amplitude and a phase, with the latter determining the relative positions of the crests and troughs of the wave. Of course, when we do a two-slit interference measurement, we are seeing the results of the relative phases of various parts of the wave function, since where two crests overlap determines where the bright fringes appear.

Unfortunately, since we haven't been able to give the full mathematical background of weak measurement theory, it will not be so clear why the measurement we describe actually does give the wave function directly. But describing the experiment and showing the results is of value in again illustrating a creative experimental method and give us an opportunity to discuss a feature of wave functions we have chosen to ignore previously. The apparatus used by researchers from the Institute of National Measurement Standards in Ottawa, Canada, is shown in Fig. 16.4. A photon is sent through a polarizer and then through a lens so that it has its wave function spread over the gray region shown. The wave function can be adjusted in shape by a mask, which, in the simplest case, is just a rectangular opening that cuts off the ends of the Gaussian form for the wave function $\Psi(x)$ in one dimension. (Recall that a Gaussian is just a simple shape, like the wave function in a harmonic oscillator; see Fig. 3.3.) A sliver of birefringent calcite rotates the polarization, but only at a specific changeable position along the $x$-axis. The photon is then in a superposition of a state of slightly different polarization at one value of $x$, and states that have the original polarization at every other $x$. The lens focuses the beam centered a small aperture. The slit picks out just the component having vanishing transverse momentum, $p_x = 0$, that is, the ray going horizontally straight through the slit. It is this component of the entire photon superposition that measures the wave function.

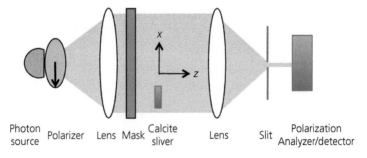

**Figure 16.4** Diagram of an apparatus to measure the photon wave function $\Psi(x)$. See text for details.

Measuring a wave function 187

The position $x$ has been encoded in the polarization. The analyzer uses the polarization of the photons in two different analysis methods, which we are not detailing here, to pick out both the amplitude and phase of the wave function. This is a good opportunity to introduce a feature of wave functions that we have previously avoided to keep the math relatively simple. There is another mathematical way to describe the amplitude and phase of the wave function. *Complex numbers* are those that can be written as $a + ib$, where $i = \sqrt{-1}$; $a$ is called the "real" part of the number, and $b$ is the "imaginary" part, because, well, the square root of $-1$ does not actually exist. Nevertheless, one can do a lot of math with complex numbers; for example, we certainly have $i^2 = -1$. Having a real and imaginary part is a bit like characterizing a number as a point in a two-dimensional space, which has the $x$ and $y$ coordinates $a$ and $b$, respectively. The wave function is complex; that is, it has a real part, denoted $\text{Re}[\Psi(x)]$, and an imaginary part, $\text{Im}[\psi(x)]$, so the total wave function is written

$$\Psi(x) = \text{Re}[\Psi(x)] + i\text{Im}[\Psi(x)]. \tag{16.2}$$

The probability of finding a particle at a position is the "magnitude" of this quantity squared, which, in complex arithmetic, is written as

$$|\Psi|^2 = \left(\text{Re}[\Psi(x)] + i\text{Im}[\Psi(x)]\text{Re}\right)\left(\text{Re}[\Psi(x)] \tag{16.3}$$
$$-i\text{Im}[\Psi(x)]\right)$$
$$= \text{Re}[\Psi(x)]^2 + \text{Im}[\Psi(x)]^2.$$

The amplitude of the wave function is given by the magnitude $|\Psi|$, while the phase, which we have described as telling us the position of the crests and troughs of the function, is given by an angle $\phi(x)$ satisfying

$$\tan \phi(x) = \frac{\text{Re}[\Psi(x)]}{\text{Im}[\Psi(x)]}. \tag{16.4}$$

If you look back at the Schrödinger equation in Eq. (1.1), you will notice an $i$ on the left side; this notation automatically means solutions can be complex—in the sense used here.

In the experiment, one way of measuring the polarization gives $\text{Re}[\Psi(x)]$, and another gives $\text{Im}[\Psi(x)]$. Fig. 16.5 shows the experimental results. (Notice that all physical quantities measured, i.e., $\text{Re}[\Psi(x)]$ and

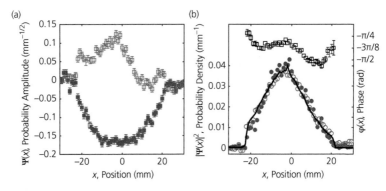

**Figure 16.5** Experimental results for the measurement of a photon wave function $\Psi(x)$. (a) $\text{Re}[\Psi(x)]$ is shown in blue (lower curve); $\text{Im}[\Psi(x)]$ is in red (upper curve). (b) The position probability (lower curve), that is, the wave function magnitude squared, as measured in two ways (red and blue dots) and as gotten by calculating $\text{Re}[\Psi(x)]^2 + \text{Im}[\Psi(x)]^2$ from the graph in (a) (solid line). The upper curve is the phase angle $\phi(x)$ from Eq. (16.4). (Reprinted by permission from Macmillan Publishers Ltd: Lundeen et al, Nature 474, 188. Copyright 2011.)

$\text{Im}[\Psi(x)]$ are themselves real. We can't actually measure $i$ itself since it is imaginary! The cut-off Gaussian is inverted, that is, has a leading minus sign, but this makes no difference in the physical quantities. The second plot shows a direct measurement of the probability of finding the photon at any $x$, as measured by counting the numbers at the various $x$ values. This agrees with the magnitude squared of the direct wave function measurement.

In another experiment, the mask is a glass plate that is put over half of the wave function area, causing a jump in the wave function that is seen in the results, which are not shown here.

Measuring a wave function had been done in previous experiments but not so directly as here. This result probably does not settle the question of whether the wave function is a matter of knowledge or a real physical quantity; but, in any case, the theory does seem to hang together amazingly well.

# Epilogue

If you have gotten this far in this book, then you should have some appreciation for the unusual character of quantum mechanics. When used with its full mathematical machinery, it is an exceedingly accurate description of nature, and yet we don't really understand it fully; its interpretation is still very controversial. The mathematical treatment of quantum mechanics used in this book, in the elementary form used, gives only a small taste of the real thing, even though I hope it gives a deeper understanding than one that depends just on words. A fuller appreciation requires a lot more: calculus, partial differential equations, and beyond. This level can be found, but still at the introductory level, in a book such as David Griffiths's *Introduction to Quantum Mechanics* or even in my own textbook with John Brehm, *Introduction to the Structure of Matter: A Course in Modern Physics*.

There are many other weird aspects of quantum mechanics that we have not treated, of course. The bibliography includes some popular books that will give more information and different insights. For example, we have not touched on much of particle physics, string theory, cosmology, or and general relativity. Books by Brian Greene and Stephen Hawking, for example, do treat those subjects without math. What we are finding is that nature seems very strange, and we have yet to grasp all of its most mysterious characteristics.

# APPENDIX

# Classical Particle Dynamics

## Newtonian Mechanics Concepts

The book is meant for readers with all kinds of backgrounds, even those having no previous introductory physics courses. For those with no previous background, the following summary will seem overwhelming—it is a whole semester course in one subsection. But the use we will have of these terms is limited and a rather intuitive understanding is all that is needed. This appendix on particle dynamics is mainly to be treated as a glossary of terms to which you can refer later as we come to use them.

*Velocity* is the rate of motion, that is, distance per unit time. A car might have a velocity of 20 miles per hour (mph). Physics usually uses meters per second (m/s) to measure velocity. Velocity is a vector, that is, it has a direction, such as 30 m/s to the north. If we just mention the size and not the direction, we call the quantity "speed"; velocity is speed in a direction. Motion in a circle has the direction constantly changing, but the speed can be constant. We might define velocity along a plus x-axis as positive, since the value of $x$ is increasing. Then, in the opposite direction, the velocity would be negative, since the value of $x$ is decreasing. In simplest terms, if a body travels at constant velocity for a distance $d$ in time $t$, then the velocity is $v = d/t$. One can consider velocity in terms of components, for example, a velocity toward the northeast can be considered a sum of a velocity to the north and a velocity to the east.

*Acceleration* is the rate of change of velocity. If I go smoothly (constant acceleration) from 20 m/s to 30 m/s in 5 seconds, my acceleration is 10 m/s divided by 5 s, or 2 m/s per second, or 2 m/s$^2$. It is also a vector and so has a direction. If I slow down, the acceleration is opposite the velocity and so can be considered negative. Suppose I start at zero velocity and have constant acceleration to a final velocity $v_f$ in a time $t$; then, my acceleration is $v_f/t$. However, my average velocity is just $v_{av} = v_f/2$ (averaging the final $v_f$ and the initial 0).

*Force* is a push or pull on a body. It is force that causes a body to accelerate. Force is vector and has a direction. Two forces of the same magnitude but in opposite directions would cancel out, leaving a net force of zero. Acceleration is the result of a net nonzero force acting on the body.

*Friction* is a force caused by surface roughness that acts when one object moves over another. Friction tends to act opposite to the direction of motion and slows an object.

*Inertia* is the resistance to change of the state of motion. A body at rest tends to stay at rest, while a body in motion tends to stay at its present velocity. It takes a force to change the velocity of the body. Inertia is a rather generic term that can refer to mass, momentum, etc.

*Mass* is a measure of the amount of material in a body. It is not the same as weight, which is a force. Weight is the effect of the force of gravity on a body and would be different on the moon than on earth. Mass is the source of a body's inertia, that is, the resistance to being accelerated by a force. Even in space away from any gravitating object, a body has mass and resists acceleration. Newton's second law is

$$F = ma. \tag{A.1}$$

The more massive a body is, so the greater its inertia, the more force is required to produce a given acceleration.

A separate property of mass is that it causes the gravitational attraction between two bodies. The force of gravity on a body on earth depends on the product of the mass of the earth and the mass of the object (as well as the distance between the objects). If you have more mass, you will weigh more on earth. The moon is less massive (and smaller) than the earth, and an astronaut on the moon weighs less there (1/6 as much), even though he or she has the same mass anywhere. When an object falls, gravity is exerting a force on it and causing it to accelerate. The rate of acceleration is the same for any object and is called $g$. On the earth, $g = 9.8$ m/s$^2$. The weight $F_W$ of an object is the force on it; this is the mass times the acceleration, according to Eq. (A.1), and so is $F_W = mg$. Einstein's general relativity theory explains gravity in terms of the distortion of space-time by mass.

Photons are massless particles of light, but because space is distorted by mass, a photon traveling close to a star will be deflected by the star.

*Momentum* for a particle with mass is the product of mass and velocity. The quantity is important because total momentum is conserved. For example, if two masses collide, they may each change velocities or directions, but the total momentum stays the same. Momentum is a vector, meaning it has a direction. One can consider components of momentum, namely, parts in the perpendicular $x$, $y$, and $z$ directions. Photons (light quanta) carry momentum, as shown in the special theory of relativity, even though they have no mass.

*Work* is the product of force and distance. If a force $F$ acts on a mass, pushing it a distance $d$, then the amount of work done is $W = Fd$. Work is important because it causes a change in energy.

*Energy* is the result of work but can appear in several different forms. Suppose the force $F$ acts alone on the body of mass $m$, starting from rest, over a distance $d$ during a time $t$. The work done is

$$W = Fd = mad = m\frac{v_f}{t}d, \tag{A.2}$$

# Appendix: Classical Particle Dynamics

where $v_f$ is the body's final velocity. Now, since the body is accelerating, $d/t$ is not the final velocity but the average velocity, $v_f/2$; so the work done is

$$W = \frac{1}{2}mv_f^2. \tag{A.3}$$

The body is said to have a **kinetic energy** of $\frac{1}{2}mv_f^2$ due to the work that was done.

Suppose we lift a body very slowly so it does not gain much kinetic energy but is raised to a height $h$. We have done work against gravity and exerted a force $mg$ through a distance $h$. The work done is

$$W = mgh, \tag{A.4}$$

and the mass has gained a *potential energy* of $mgh$. Note that if I now drop the object, gravity accelerates it, so it gains kinetic energy. The potential energy has been converted to kinetic energy, and the total energy has been conserved, just transformed from one kind to another. Wait a minute! Where did the original work energy come from? From my muscles; I converted *chemical energy* into a force exerted over a distance in lifting the body against gravity. The object stopped when it hit the ground—the kinetic energy disappeared and became *thermal energy*, which is the total of kinetic and potential energies of the internal random motion of the molecules of the system. Temperature is often a measure of the average kinetic energy in this random motion. Heat is thermal energy being transferred from a hot object to a colder object. Frictional forces do work that generates heat and causes a loss of kinetic and potential motion to thermal energy.

There are other forms of potential energy besides gravitational. Suppose I stretch a spring. The spring pulls back with more force the more I stretch it. The force depends on the distance $x$ it has been stretched; it is given by $F_x = kx$, where $k$ is the "spring constant." The total work done in stretching it a total of $d$ distance is $F_{av}d$ where $F_{av} = kd/2$ is the average force (the force is changing as I stretch it). So, the total elastic potential energy is

$$PE = \frac{1}{2}kd^2. \tag{A.5}$$

If I attach a mass to the spring, stretch it out, and let it go, it will oscillate, changing potential energy to kinetic energy and back again (we neglect friction). This is easier to see in the pendulum shown in Fig. A.1, but I also show the spring oscillating around equilibrium. The potential and kinetic energies add to a constant value (if we neglect the effects of friction), and the pendulum will oscillate forever.

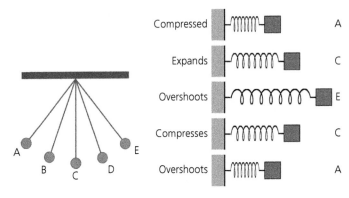

**Figure A.1** *Left*: A pendulum is lifted to point A and released. At B it has converted some of its potential energy to kinetic energy. At C all the energy is kinetic energy. At D it has some potential energy and some kinetic energy, but, at E, it is again all potential energy. The cycle then repeats in the other direction. Oscillation would continue forever, except that friction ultimately converts all the energy to thermal energy. *Right*, a spring undergoing oscillation. When fully compressed or fully extended (A and E, respectively) the energy is all potential energy, but, in-between, it is part kinetic energy and part potential energy, except at the center point C, where the spring is neither compressed nor extended. Then, the energy is all kinetic energy.

We will need to see plots of potential energy (often called $V(x)$, where the position of the particle is $x$) of the mass on a spring, as well as for other systems. Study Fig. A.2.

This has letters corresponding to the positions on the spring (or pendulum). The total energy is taken as $E = 9$ J.[1] When the mass on the spring is at position A, the spring is compressed to the left (or the pendulum is at its highest point on the left). The kinetic energy at that point is zero; perhaps we are holding the mass at that point, about to let it go. The potential energy at any point $x$ is, as we saw in Eq. (A.5),

$$V(x) = \frac{1}{2}kx^2. \qquad (A.6)$$

---

[1] The energy of everyday-sized objects is usually measured in joules. A fast pitcher throws a baseball with 140 joules of kinetic energy. But the "J" in our energy could stand for some other appropriate energy unit too. An atom oscillator might have only $10^{-31}$ joules of potential energy.

Appendix: Classical Particle Dynamics 195

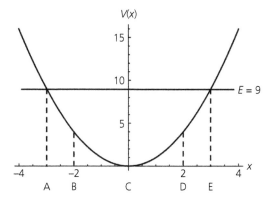

**Figure A.2** The plot of potential energy $V(x)$ for a mass on a spring or a pendulum. The straight line across is the value of the total energy ($E = 9$ J) for all positions. The potential energy varies quadratically with x, that is, it grows as $x^2$ for either positive or negative positions. The letters correspond to the positions of the mass on the spring or the pendulum in Fig. A.1.

Then, at point $x = -3$ m (m is a unit of length, say, meters), with, in our example, $k = 2$ J/m², we have

$$V(-3) = \frac{1}{2} 2 \times (-3)^2 = 9 \text{ J}. \quad (A.7)$$

We now let the mass go, and the compressed spring pushes the mass to accelerate from zero to velocity toward the right. The potential energy gets less while

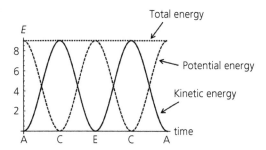

**Figure A.3** The plot of potential energy, kinetic energy, and total energy of mass on a spring, as a function of time. The total energy stays constant. The letters show the times corresponding to the various positions shown in Fig. A.1; italic E stands for energy.

the kinetic increases, with the total staying at 9 J. At point B, $x = -2$, we have

$$V(-2) = \frac{1}{2} 2 \times (-2)^2 = 4 \text{ J}. \qquad (A.8)$$

Thus, the kinetic energy is $K = 9 - 4 = 5$ J. The mass's velocity increases until it reaches point C, where $x = 0$, and $V(0) = 0$. The spring is neither compressed nor extended now. The energy is now *all* kinetic: $K = 9$ J. As the mass's momentum keeps it moving to the right, it starts compressing the spring, and the potential energy increases again. At point D ($x = 2$ m), we again have $V = 4$ J, and $K = 5$ J; and, at E, $V = 9$ J, with $K = 0$ J. The mass reverses direction and repeats the process in the opposite direction. A plot of kinetic and potential energies versus *time* is given in Fig. A.3.

# Bibliography

[1] William J. Mullin, William J. Gerace, Jose P. Mestre, Shelley L. Velleman, *Fundamentals of Sound with Applications to Speech and Hearing* (second edition), Off the Common Press (2016).

[2] John Bell, *Speakable and Unspeakable in Quantum Mechanics*, Cambridge University Press (1987).

[3] David Lindley, *Where Does The Weirdness Go? Why Quantum Mechanics Is Strange, But Not As Strange As You Think*, Basic Books (1997); *Uncertainty: Einstein, Heisenberg, Bohr, and the Struggle for the Soul of Science*, Anchor Books (2008).

[4] Anton Zeilinger, *Dance of the Photons: From Einstein to Quantum Teleportation* Ferrar, Straus and Giroux (2010).

[5] Nick Herbert, *Quantum Reality: Beyond the New Physics*, Anchor Books (1985).

[6] Brian Greene, *The Fabric of the Cosmos: Space, Time, and the Texture of Reality*, Vintage (2005); *The Elegant Universe: Superstrings, Hidden Dimensions, and the Quest for the Ultimate Theory* (second edition), W. W. Norton & Company (2010); *The Hidden Reality: Parallel Universes and the Deep Laws of the Cosmos*, Vintage (2011).

[7] Nicolas Gisin, *Quantum Chance: Nonlocality, Teleportation and Other Quantum Marvels*, Copernicus (2014).

[8] Franck Laloë, *Do We Really Understand Quantum Mechanics?*, Cambridge University Press (2011).

[9] M. Arndt and K. Hornberger, "Insight review: Testing the limits of quantum mechanical superpositions," Nat. Phys. 10, 271–277 (2014).

[10] A. Einstein, B. Podolsky, and N. Rosen, "Can quantum mechanical description of physical reality be considered complete?" Phys. Rev. 47, 777–780 (1935).

[11] D. Bohm, "A suggested interpretation of the quantum theory in terms of hidden variables," Phys. Rev. 85, 166–179 and 180–193 (1952).

[12] D. Bohm and J. Bub, "A proposed solution to the measurement problem in quantum mechanics by hidden variable theory," Rev. Mod. Phys. 38, 453–469 (1966).

[13] S. Freedman and J. Clauser, "Experimental tests of local hidden variable theories," Phys. Rev. Lett., 28, 938–941 (1972).

[14] E. S. Fry and R. C. Thompson, "Experimental test of local hidden-variable theories," Phys. Rev. Lett. 37, 465–468 (1976).

[15] A. Aspect, P. Grangier, and G. Roger, "Experimental tests of realistic local theories via Bell's theorem," Phys. Rev. Lett., 47, 460–463 (1981).

[16] F. Kaiser, T. Coudreau, P. Milman, D. B. Ostrowsky, and S. Tanzilli, "Entanglement-enabled delayed-choice experiment," Science 338, 637 (2012).

[17] N. D. Mermin, "Bringing home the atomic world: Quantum mysteries for anybody," Am. J. Phys. 49, 940–943 (1981); "Quantum mysteries revisited," Am. J. Phys. 58, 731–734 (1990).
[18] J. F. Clauser, M. A. Horn, A. Shimony, and R. A. Holt, "Proposed test to test hidden-variables theories," Phys. Rev. Lett. 23, 880–884 (1969).
[19] H. Everett III, "Relative state formulation of quantum mechanics," Rev. Mod. Phys. 29, 454–462 (1957).
[20] A. J. Leggett, "Testing the limits of quantum mechanics: Motivation, state of play, prospects," J. Phys.: Condens. Matter 14, R415 (2002).
[21] S. Haroche, "Nobel lecture: Controlling photons in a box and exploring the quantum to classical boundary," Rev. Mod. Phys. 85, 1083–1102 (2013); "The secrets of my prizewinning research," Nature 490, 311 (2012).
[22] G. C. Ghirardi, A. Rimini, and T. Weber, "Markov processes in Hilbert space and continuous spontaneous localization of systems of identical particles," Phys. Rev. D 34, 470 (1986).
[23] P. Pearle, "Combining stochastic dynamical state-vector reduction with spontaneous localization," Phys. Rev. A 39, 2277–2289 (1989); F. Laloë, W. J. Mullin, and P. Pearle, "Heating of trapped ultracold atoms by collapse dynamics," Phys. Rev. A 90, 052119 (2014).
[24] C. A. Fuchs, N. D. Mermin, and R. Schack, "An introduction to QBism with an application to the locality of quantum mechanics," Am. J. Phys. 82, 749 (2014); http://arxiv.org/abs/1311.5253; C. A. Fuchs, "QBism, the perimeter of quantum Bayesianism," http://arxiv.org/abs/1003.5209.
[25] L. Hardy, "Are quantum states real," Int. J. Mod. Phys. 27, 1345012 (2013).
[26] M. F. Pusey, J. Barrett, and T. Rudolph, "On the reality of the quantum state," Nat. Phys. 8, 476–479 (2012).
[27] J. Emerson, D. Serbin, C. Sutherland, and V. Vetch, "The whole is greater than the sum of the parts: On the possibility of purely statistical interpretation of quantum theory," http://arxiv.org/abs/1312.1345.
[28] M. Ringbauer, B. Duffus, C. Branciard, E. G. Cavalcanti, A. G. White, and A. Fedrizzi, "Measurements on the reality of the wavefunction," Nat. Phys. 11, 249254 (2015).
[29] R. Galkiewicz and R. Hallock "Observation of persistent currents in a saturated superfluid film," Phys. Rev. Lett. 33, 1073–1076 (1974).
[30] K. B. Davis, M.-O. Mewes, M. R. Andrews, N. J. van Druten, D. S. Durfee, D. M. Kurn, and W. Ketterle, "Bose-Einstein condensation in a gas of sodium atoms," Phys. Rev. Lett. 75, 3969–3973 (1995).
[31] M. H. Anderson, J. R. Ensher, M. R. Matthews, C. E. Wieman, and E. A. Cornell, "Observation of Bose-Einstein condensation in a dilute atomic vapor," Science 269, 198 (1995).

# Bibliography 199

[32] M. R. Andrews, C. G. Townsend, H.-J. Miesner, D. S. Durfee, D. M. Kurn, and W. Ketterle, "Observation of interference between two Bose condensates," Science 275, 637–641 (1997).

[33] H. Müntinga, H. Ahlers, M. Krutzik, A. Wenzlawski, S. Arnold, D. Becker, K. Bongs, H. Dittus, H. Duncker, N. Gaaloul, C. Gherasim, E. Giese, C. Grzeschik, T. W. Hänsch, O. Hellmig, W. Herr, S. Herrmann, E. Kajari, S. Kleinert, C. Lämmerzahl, W. Lewoczko-Adamczyk, J. Malcolm, N. Meyer, R. Nolte, A. Peters, M. Popp, J. Reichel, A. Roura, J. Rudolph, M. Schiemangk, M. Schneider, S. T. Seidel, K. Sengstock, V. Tamma, T. Valenzuela, A. Vogel, R. Walser, T. Wendrich, P. Windpassinger, W. Zeller, T. van Zoest, W. Ertmer, W. P. Schleich, and E. M. Rasel, "Interferometry with Bose-Einstein condensates in microgravity," Phys. Rev. Lett. 110, 093602 (2013).

[34] W. J. Mullin and F. Laloë, "Creation of NOON states from double Fockstate/Bose-Einstein condensates," J. Low Temp. Phys. 162, 250257 (2011).

[35] W. G. Unruh, "Experimental black-hole evaporation," Phys. Rev. Lett. 46, 1351–1355 (1981).

[36] S. W. Hawking, "Particle creation by black holes," Commun. Math. Phys. 43, 199 (1975).

[37] S. Hawking, (1988). *A Brief History of Time*, Bantam Books (1988).

[38] D. Boiron, A. Fabbri, P.-É. Larré, N. Pavloff, C. I. Westbrook, and P. Zi "Quantum signature of analog Hawking radiation in momentum space," Phys. Rev. Lett. 115, 025301 (2015).

[39] W. M. Itano, D. J. Heinzen, J. J. Bollinger, and D. J. Wineland, "Quantum Zeno effect" Phys. Rev. A 41, 2295–2300 (1990).

[40] P. Kwiat, H. Weinfurter, and A. Zeilinger, "Quantum seeing in the dark," Sci. Am., 275, 72–78 (1996).

[41] A. Elitzur and L. Vaidman, "Quantum mechanical interaction-free measurements," Found. Phys. 23, 987–997 (1993).

[42] P. Kwiat, H. Weinfurter, T. Herzog, and A. Zeilinger, "Interaction-free measurement," Phys. Rev. Lett., 74, 4763–4766 (1995); G. B. Lemos, V. Borish, G. D. Cole, S. Ramelow, R. Lapkiewicz, and A. Zeilinger, "Quantum imaging with undetected photons," Nature 512, 409 (2014).

[43] M. Scully, B.-G. Englert and H. Walther, "Quantum optical tests of complementarity," Nature 351, 111–116 (1991).

[44] R. Mir, J. Lundeen, M. Mitchell, A. Steinberg, J. Garretson, and H. Wiseman, "A double-slit which-way experiment on the complementarity-uncertainty debate," New J. Phys. 9, 287 (2007).

[45] R. Hillmer and P. Kwiat, "A do-it-yourself quantum eraser," Sci. Am., 296, 90–95 (2007).

[46] F. Colegrove Jr., L. Schearer, and G. Walters, "Polarization of $^3$He gas by optical pumping," Phys. Rev. 132, 2561–2572 (1963).

[47] C. Hong, Z. Ou, and L. Mandel, "Measurement of subpicosecond time intervals between two photons by interference," Phys. Rev. Lett. 59, 2044–2046 (1987).
[48] A. Kaufman, B. Lester, C. Reynolds, M. Wall, M. Foss-Feig, K. Hazzard, A. Rey, and C. Regal, "Two-particle quantum interference in tunnel-coupled optical tweezers," Science 345, 306–309 (2014).
[49] R. Hanbury Brown and R. Twiss, "Correlation between photons in two coherent beams of light," Nature 177, 2729 (1956).
[50] T. Jeltes, J. McNamara, W. Hogervorst, W. Vassen, V. Krachmalnicoff, M. Schellekens, A. Perrin, H. Chang, D. Boiron, A. Aspect, and C. Westbrook. "Comparison of the Hanbury Brown-Twiss effect for bosons and fermions," Nature 445, 402–405 (2007).
[51] David Griffiths, *Introduction to Elementary Particles* (second edition), Wiley-VCH (2008); *Introduction to Quantum Mechanics* (second edition), Addison Wesley (2004).
[52] Richard P. Feynman, *QED: The Strange Theory of Light and Matter*, Princeton University Press (1985).
[53] B. P. Abbott et al (LIGO Scientific Collaboration and Virgo Collaboration, "Observation of gravitational waves from a binary black hole merger," Phys Rev. Lett. 116, 061102 (2016).
[54] A. Lambrecht, "The Casimir effect: A force from nothing," Phys. World, 15, 29 (2002).
[55] S. Lamoreaux, "Demonstration of the Casimir force in the 0.6 to 6 mm range," Phys. Rev. Lett. 78, 5–8 (1997).
[56] Robert Oerter, *The Theory of Almost Everything: The Standard Model, the Unsung Triumph of Modern Physics*, Plume (2006).
[57] X.-S. Ma, T. Herbst, T. Scheidl, D. Wang, S. Kropatschek, W. Naylor, B. Wittmann, A. Mech, J. Kofler, E. Anisimova, V. Makarov, T. Jennewein, R. Ursin, and A. Zeilinger, "Quantum teleportation over 143 kilometres using active feed-forward between two Canary Islands," Nature 489, 269273 (2012); http://arxiv.org/abs/1205.3909.
[58] X.-H. Bao, X.-F. Xu, C.-M. Li, Z.-S. Yuan, C.-Y. Lu, and J.-W. Pan, "Quantum teleportation between remote atomic-ensemble quantum memories," http://arxiv.org/abs/1211.2892; http://www.technologyreview.com/s/507531/first-teleportation-from-one-macroscopic-object-to-another/.
[59] N. David Mermin, *Quantum Computer Science: An Introduction*, Cambridge University Press (2007).
[60] Y. Aharonov, D. Albert, and L. Vaidman, "How the result of a measurement of a component of the spin of a spin-1/2 particle can turn out to be 100," Phys. Rev. Lett. 60, 1351–1354 (1988).

[61] I. Duck, P. Stevenson, and E. Sudarshan, "The sense in which a weak measurement of a spin-1/2 particle's spin component yields a value 100," Phys. Rev. D 40, 2112–2117 (1989).

[62] S. Kocsis, B. Braverman, S. Ravets, M. Stevens, R. Mirin, L. Shalm, and A. Steinberg, "Observing the average trajectories of single photons in a two-slit interferometer," Science 332, 1170–1173 (2011).

[63] J. Lundeen, B. Sutherland, A. Patel, C. Stewart, and C. Bamber, "Direct measurement of the quantum wavefunction," Nature 474, 188–191 (2011).

[64] Banesh Hoffmann, The Strange Storey of the Quantum, Dover Publication (1947). Courtesy of the Banesh Hoffmann Estate

[65] Henry P. Stapp, "Bell's Theorem and World Process," Nuovo Cimento, 29B, 2, 270 (1975).

# Index

α-particle 27, 142
β-particle 142
γ-particle 142
γ-ray 12
π meson 159
$^3$He 89, 111, 126
$^3$He gas 114, 116, 118, 119
$^3$He liquid 117
$^4$He 89, 126
$^4$He gas 114
$^4$He liquid 93, 100, 118, 119
xyz coordinate system 35

## A

absolute zero 86, 93, 95
acceleration 191
Achilles 101
Adleman, Leonard 176
alkali atoms 86, 87
alkali gases 95
ammonia 27, 28
amplitude 6
Anderson, Philip 164
angular momentum 34, 37
antibunching 125
antiferromagnet 120
antinode 6
antisymmetric wave function 91, 112, 116
Arndt, Marcus 31
Aspect, Alain 65
asymptotic freedom 160
atom laser 96
atomic clock 96
atomic physics 2

## B

baryon 159
Bayes, Thomas 77
Bayesian interpretation 70, 77
beam splitter 49, 50, 106
Bell experiments 65
Bell's theorem 44, 56, 63, 80
Bell, John 44, 63, 64
Bell state 167, 168, 170
Bethe, Hans 155
Big Bang 161
binary number 172

binary pulsar 158
birefringent crystal 183
black hole 115
black-body radiation 11
Bohm, David 45, 46, 56
Bohm interpretation 45, 80, 81, 182
Bohr, Niels 22, 31, 53
Boltzmann constant 114
Born, Max 14
Bose gas 114
Bose principle 92, 94
Bose, Satyendra Nath 85
Bose–Einstein condensate 67, 71
Bose–Einstein condensation 85, 86, 92, 93, 95
Bose–Einstein statistics 91, 111
boson 90, 111, 144
bottom quark 159
box potential 12
Brossel, Jean 117
bunching 125

## C

carbon-60 31, 70
carbon dating 143
Caron, Christopher ix
Casimir, Hendrik 157
Casmir force 157
cavity 130, 145
Celsius scale 86
centigrade scale 87
CERN 33
change of basis 41, 133, 165
charmed quark 159
chemical behavior 89
chirality 104
CHSH inequality 64
classical mechanics 1
classical physics viii
Clauser, John 64, 65
closed shell 91
Cohen-Tannouji, Claude 117
Colegrove, Jr., Forrest 117
collapse of the wave function 26, 31, 40, 68, 76
complementarity 3, 10
complete set 168
complex numbers 187

condensate interference 96–99
conjugate variable 47
continuous spontaneous localization 76
Cooper pair 163
Copenhagen interpretation 26, 31, 66, 69, 76, 84
Cornell, Eric 97
correlation function 124
correspondence principle 22
Coulomb unit 141
Coulomb's law 141
counterfactual computing 106
coupling constant 143
creeping film 95
critical velocity 95
cycle 6

D

D-Wave Systems 180
Davisson, Clinton 10
de Broglie, Louis 10
decoherence 31, 71, 74
delayed choice 53
Deutch, John 175
Deutch's problem 175, 177
dice 15
dilute gases 95
Dirac, Paul 167
Dirac notation 167
direct term 124
double-well potential 23–25, 32
down quark 159

E

Einstein, Albert viii, 10, 42, 85
Einstein, Podolsky, and Rosen 42, 46
elastic potential energy 193
electromagnetic force 140
electromagnetic interaction 141, 154, 156
electromagnetic radiation 10
electron 34, 39, 72, 90, 112, 161
electroweak interaction 161
elements of reality 43, 44, 64
Elitzur, Avshalom 106
energy 1, 3, 11, 192
energy conservation 19
energy gap 94, 95
energy, total 17, 23
Englert, Berthold-Georg 130
entangled state 32, 40, 53, 54, 174, 178
entanglement 32
epistemic interpretation 66, 77, 80

EPR 42, 46, 48
EV effect 106
Everett interpretation 70
exchange force 140
exchange term 113, 123

F

factoring 176
Fahrenheit scale 86
femtometers 89
Fermi gas 114
Fermi–Dirac statistics 90, 112
fermion 90, 111
ferromagnet 120
Feynman, Richard 150, 182
Feynman diagram 145, 150, 153, 156
field 145
fine structure constant 144
first harmonic 8
force 18, 191
fountain effect 86
Fourier analysis 8
Freedman, Stuart 65
frequency 6, 11, 18
friction 191
frictionless flow 93
Fry, Edward S. 65
Fuchs, Christopher 78
fundamental frequency 8

G

g-factor 144
Gaussian 20, 186
Gell-Mann, Murray 159
general theory of relativity 1, 141, 158, 192
Gerlach, Walther 34
Germer, Lester 10
Ghirardi, Gian Carlo 76
Glashow, Sheldon 161
Glauber, Roy 123
gluino 163
gluon 34, 91, 159
Grangier, Philippe 65
gravitational force 140, 192
gravitational potential energy 193
graviton 158
gravity 1, 140, 158
gravity waves 158
Grover, Lov 176
Grover's search algorithm 176
GRWP theory 76

## H

hadron 159
half-life 67, 142
Hallock, Robert 95
Hanbury Brown, Robert 122
Hanbury Brown–Twiss experiment 121
Hardy, Lucian 82
harmonic oscillator 16
harmonic potential 88
harmonic series 8
Haroche, Serge 71, 117
Hawking, Stephen 100, 157
Hawking radiation 100, 157
Heisenberg uncertainty principle vii, 11, 20, 46
Heisenberg, Werner 10
helium 27, 89, 91
helium gas 114, 116, 118, 119
helium liquid 2, 85, 86, 93, 100, 117–119
hertz 6
hidden variables 43, 44, 56, 61, 81
Higgs boson 33, 91, 162
Holt, Richard 64
Hong, Chung-Ki 120
Hong–Ou–Mandel effect 120
Horne, Michael 64
Hulse, Russell 158
hydrogen 89
hydrogen states 90
hyperfine states 87

## I

Ice Cube 143
ideal gas law 114
imaginary part 187
incoherent source 75
indistinguishable particles 32
inertia 192
information leakage 73
input register 173, 177
interaction vertex 156
interaction-free measurement 106, 107, 109
interference 5, 30, 67
interference, constructive 6, 23, 49
interference, destructive 6, 23, 49
interferometer 49, 71
interpretation of quantum mechanics 2, 26, 31, 45, 66, 69, 70, 76, 77, 80, 84
isotope 89, 142

## J

Joule 18, 194

## K

Kastler, Alfred 116
Kelvin scale 86
Ketterle, Wolfgang 97
kinetic energy 12–14, 18, 23, 193

## L

Laloë, Franck ix
Lamb, Willis 155
Lamb shift 155
Lamoreaux, Steve 157
Large Hadron Collider 33, 162
laser 92
laser cooling 88
laser trap 87, 88, 96
Leggett, Anthony 70
lepton 162
LIGO 158
Lindley, David 68
local realism 44
London, Fritz 93
low temperature 67, 85

## M

Mach–Zehnder interferometer 49, 52, 82, 98, 107
magnetic resonance imaging 34, 100, 118
magnetic trap 87, 88, 96
magnets 33
magnitude 187
Malus law 51, 105
Mandel, Leonard 120
mass 1, 192
matter waves 10, 96
measurement amplification 182
medium motion 4
medium, elastic 4
Mermin, N. David 56, 180
Mermin machine 56, 57
meson 159
Michelson interferometer 122
mixed state 75
momentum 1, 3, 10, 11, 14, 192
MRI 34, 100, 118

## N

Ne'eman, Yuval 159
neutrino 143, 162
neutron 34, 90, 159
neutron star 115, 158
Newton, Isaac 29
Newton's law of gravity 140

# Index

Newton's laws 1, 18
Newton's second law 192
Newtonian mechanics viii, 191
NMR 34, 152
node 6, 22
nonlocality viii, 43, 48, 56
NOT operator 173
nuclear magnetic resonance 34, 152

## O

Onnes, Kamerlingh 85
ontic interpretation 66, 76, 77, 80
optical coherence 123
optical potential 88
optical pumping 92, 116
optical trap 96
optically active 104
Ou, Zhe-Yu 120
output register 173, 177
Ozawa uncertainty principle 12

## P

pair annihilation 145
pair production 145
particle 10
particle behavior 3, 30, 31, 50, 53, 54
particle zoo 162
particle–wave duality vii, 30
path integral method 182
Pauli pressure 115
Pauli principle 91, 112, 115
PBR theorem 83
Pearle, Philip 76
pendulum 16
period 6
period finding 176
phase angle 187
photino 163
photoelectric effect 11
photon vii, 10, 19, 34, 49, 72, 91, 192
photon exchange 153
picometers 89
Planck's constant 11, 34
Planck, Max 10
Podolsky, Boris 42
polarization 33, 50, 104, 137
polarization displacer 184
polarization, horizontal 39, 51
polarization, vertical 39, 51
polarizer 51
polarizing beam splitter 51
polaron 163
positron 39

potential energy 12, 16, 18, 23, 24, 193
probability 14, 15, 21, 23, 28, 66, 76, 77, 102
projective measurement 181
proton 90, 159
pulsar 158

## Q

QBists 77
quanta 10
quantization of energy 19
quantization of spin 37
quantum Bayesianism 70, 77
quantum computing 3, 33, 171, 172
quantum crystal 93
quantum electrodynamics 150
quantum eraser 130
quantum parallelism 174
quantum pendulum states 21
quantum tunneling vii, 16, 21, 24, 27, 93
quark 34, 90, 91, 159
quasiparticle 163
qubit 171, 173

## R

Rabi, Isidor 152
Rabi frequency 102
Rabi oscillations 102, 152
radioactive atom 67
radioactivity 142
real part 187
reality of the wave function 79
reflection 5
refraction 183
relativity theory 42
relaxation time 118
renormalization 155
Rimini, Alberto 76
Rivest, Ron 176
Roger, Gérard 65
Rosen, Nathan 42
rotational symmetry 42
RSA encryption 176

## S

Salam, Abdus 161
Schearer, Laird 117
Schrödinger, Erwin vii, 79
Schrödinger equation viii, 3, 10, 19, 45, 76
Schrödinger's cat 31, 55, 66, 71, 78, 99
Scully, Marian 130
selectron 163

self-energy diagram 154
shadowgram 97
Shami, Adi 176
Shimony, Abner 64
Shor, Peter 176
Shor's period-finding algorithm 176
singlet state 40, 79
sinusoidal wave 14
slope 18
spectrum, continuous 9
spectrum, discrete 8
speed 191
spin 32, 33
spin 1/2 34, 37, 90, 115
spin polarization 116
spontaneous parametric down-conversion 53
spooky action at a distance viii, 42, 65
spring constant 193
squark 163
SQUID 70
Standard Model 159, 162
standing wave 5, 7, 12, 14, 23
Stern, Otto 34
Stern–Gerlach apparatus 35, 57, 67
Stern–Gerlach apparatus, double 38
stimulated emission 92
strange quark 159
string theory 1, 37
strong nuclear force 140, 142
superconducting quantum interference device 70
superconductor 85, 100
superfluid helium 85, 93, 118
superposition vii, 5, 8, 26, 49, 69, 71, 77
supersymmetry 163
symmetric wave function 92, 112
symmetry 24

T

Taylor, Joseph 158
teleportation 33, 165, 175
temperature 86
thermal conductivity 118
Thompson, Randall C. 65
time–energy uncertainty principle 27, 151
top quark 159
transport processes 118
triplet state 40
turning points 18, 21
Twiss, Richard 122

two-slit experiment 28, 29, 45, 128
two-slit trajectories 182

U

uncertainty principle, Heisenberg vii, 9, 11, 20, 46
uncertainty principle, Ozawa 12
uncertainty principle, time–energy 27, 151
up quark 159

V

vacancy 28
vacuum 144
vacuum polarization 155, 157, 163
Vaidman, Lev 106
van der Waals interaction 114
vector 191
velocity 191
virial coefficient 114
virtual bosons 144
virtual particles 140
viscosity 119
vortices 99

W

$W$ boson 34, 91, 160
Walters, Geoffrey 117
Walther, Herbert 130
wave 10
wave behavior vii, 3, 30, 31, 50, 53, 54
wave dynamics 3
wave function 2, 3, 12, 14, 16, 19, 66, 67, 69–73, 75–84
wave function, antisymmetric 24, 26
wave function, symmetric 24, 26, 92
wave motion 4
wave velocity 6
wave, impulsive 4
wave, oscillatory 4
wave, sinusoidal 5, 10
wave, standing 5, 7, 12, 23
wave, transverse 4
wavelength vii, 5, 11, 14
weak measurement 181
weak nuclear interaction 140, 142, 160
Weber, Tullio 76
weight 192
Weinberg, Steven 161
Wheeler, John 53
which-way information 130
white dwarf 115
Wieman, Carl 97
Wigner, Eugene 68

Wigner's friend  69, 78
Wineland, David  104
work  192

**Y**

Young, Thomas  29
Yukawa, Hideki  159

**Z**

Z boson  34, 91, 160
Zeilinger, Anton  106, 170
Zeno effect  101, 104
Zeno of Elea  101
zero-point energy  14, 20, 93
Zweig, George  159